God, Probability, and Life after Death

God, Probability, and Life after Death

An Argument for Human Resurrection

William Hunt

LEXINGTON BOOKS
Lanham • Boulder • New York • London

Published by Lexington Books
An imprint of The Rowman & Littlefield Publishing Group, Inc.
4501 Forbes Boulevard, Suite 200, Lanham, Maryland 20706
www.rowman.com

Unit A, Whitacre Mews, 26-34 Stannary Street, London SE11 4AB

British Library Cataloguing in Publication Information Available

Library of Congress Cataloging-in-Publication Data Available

ISBN 978-1-4985-2606-7 (cloth : alk. paper)
ISBN 978-1-4985-2607-4 (electronic)

∞™ The paper used in this publication meets the minimum requirements of American National Standard for Information Sciences—Permanence of Paper for Printed Library Materials, ANSI/NISO Z39.48-1992.

Printed in the United States of America

To Helen, Nicola, Robert, Peter, Amber, and Liberty

Contents

Preface

One Sunday a year or so ago, after some stirring hymns, I sat down in my usual pew in St John's the Evangelist Church in Blackheath, London, and the Reverend Anne Williamson stepped up to the lectern to address the congregation. Her subject was human resurrection and the promise made to mankind by Jesus. She emphasised the essential materialist view of the Christian faith on this matter; that is, bodily resurrection rather than the oft thought spiritual resurrection, with the soul fluttering off to Heaven. I was both captivated by her presentation and perplexed by the thought of a dead person, whose body had putrefied and returned to dust, rising up to live once again. Such thoughts haunted me in the following days and weeks and I became determined to explore my thoughts on this issue and try to find an explanation for human resurrection other than *God moves in mysterious ways*. The following chapters represent my journey of exploration in this fascinating domain, where I engaged with that question that nearly everyone asks at some point in their life: "Is there life after death?"

Acknowledgments

Heythrop College, University of London, United Kingdom

I should also like to express my thanks to the following who have given me advice and support in the writing of this book: Michael Donovan, Professor Michael Barnes, Professor Donald Gillies, Professor Colin Howson, Professor Gwen Griffith-Dickson, Dr. Andris Abakuks, Dr Jenifer Bailey, Ronald Barkham and Chris Romer.

I should particularly like to thank John Cartlidge for acting as an evaluator in helping to determine the vicarious probability values in four of the chapters.

Acknowledgments

Introduction

The human resurrection hypothesis is not confined to the Christian religion; most religions engage with the question in some way and typically they have a general eschatology and sometimes a personal one in their doctrines. My approach herein has been to present a personal resurrection hypothesis; that is, individual resurrection upon death, although the arguments presented do have application to the resurrection of the whole of mankind in a general eschatological hypothesis.

There are a very large number of different religions, past and present, with differing eschatological doctrines within them, but I limit myself to a brief comment on five world-religions and their resurrection principles.

Hinduism, is probably the oldest and most complex religion in the world with both atheistic and theistic notions within its differing systems. Hinduism has a personal rather than general resurrection hypothesis based upon the notions of *karma* and *samsara*. Karma is a cause and effect concept whereby the quality of the afterlife reflects the deeds done in Earthly life; and samsara is the repetition of birth, life, and death. Re-births can be in differing planes; Heavens, Hells, and on Earth, with life forms on Earth being varied according to the individual's karma. Generally, Hinduism embraces the concept of a soul that can either exist as is, or acquire another material existence; the former being the ideal post-mortem existence.

Buddhism sprang from Hinduism and is as diverse in its beliefs and practices. Though not theistic, some Buddhist traditions hold to the view of endless life, either by reincarnation or nirvana, a perfected state of mind free from repeating reincarnations. However, not all Buddhist tradition holds to the idea of rebirth.

Like Hinduism, Buddhism has a personal eschatology rather than a general one, but unlike Hinduism it rejects the notion of a human soul in favor of a conjunction of physicality and mental-states; mental-states seemingly being able to survive without physicality.

Zoroastrianism is an ancient religion of the Persian empire with a resurrection philosophy that had a strong influence on the later Abrahamic religions of Judaism, Christianity, and Islam. It has both a personal and general eschatology and is a dualistic (body and soul) religion. The notions of Heaven, Hell, and bodily resurrection, common to Christianity, originate in Zoroastrianism. Like the Judaic, Christian, and Islamic religions, the general eschatology of Zoroastrianism involves a final confrontation between God and evil resulting in a general resurrection and judgement of mankind.

There is no systematic personal eschatology in Judaism despite references to Heaven and Hell in the literature. There is, however, Sheol or "pit for the dead," a later interpretation in Judaism, where the dead await judgement and general bodily resurrection with the reestablishment of Jerusalem.

Christianity's eschatology is like Judaism with more emphasis on a general rather than personal resurrection, although a personal resurrection scenario can be inferred from the New Testament. Indeed, Hades, like Sheol, is an intermediary holding-world pending the general resurrection. This two-phase event involves the return of Christ in the new Jerusalem, the destruction of evil, and the resurrection of all mankind prior to judgement. Phase one of the general resurrection is for the very righteous; phase two is for everyone else who are then judged: the righteous survive and the wicked are destroyed. Both Judaism and Christianity echo the eschatology of Zoroastrianism.

Islam has a clear personal eschatology based upon dualism, where the righteous soul survives death to live comfortably in their supernaturally converted graves pending the general physical resurrection; note the similarity to Sheol. The same state of affairs occurs for the unrighteous, but their graves are a place of suffering. The general resurrection again reflects the eschatology of Zoroastrianism; a bodily resurrection in paradise.

The writing of this book was a personal journey of discovery, as before commencing writing I was not certain what result my analysis would yield. This is because it was a cumulative probabilistic analysis; that is, consideration of the probable truth of a set of propositions relative to relevant evidence resulting in the formation of a degree of belief in their truth. Cumulative, because the degree of truth of each proposition considered impinges on the degree of belief in the truth of the next proposition, and so on until the set of propositions is exhausted and a result emerges. The central proposition under consideration is that of human resurrection; namely, *dead person will be resurrected.* My approach was primarily from the proposition *God exists* through the proposition *Jesus died and was Resurrected* followed by the proposition *there are perceived conscious visitations to Heaven or other supernatural realms by persons whilst in a near-death state* finalizing with the proposition *living persons experience apparitions of dead persons*. The result that emerged surprised me; that is, the probability of the proposition *dead person will be resurrected*, in terms of a degree of belief in that proposition was high, enabling me to claim that human resurrection is a distinct possibility.

In the personal resurrection hypothesis presented herein, I have placed God at the center of the argument. Initially, my reason for this choice emerged for reasons stated in the Preface above and, of course, the will of God is fundamental to the understanding of human resurrection in Judaism, Christianity, and Islam. However, I have opted to take a more philosophical rather than theological approach; that is, primarily based upon my understanding of the attributes of God. Notwithstanding, because I engage with the evidence for the resurrection of Jesus in my analysis, my overall argument is seen, to some extent, through the lens of Christianity.

Human resurrection, I argue, is what we would expect from a being possessing such attributes in terms of the desire for it and the wherewithal to ensure it. Volumes have been written on the fascinating topic of the attributes of God, some of which I have referenced and, given the force of these attributes to my overall argument, I have allocated considerable space to it. However, placing God at the center of the argument does presuppose God exists and I have allocated even more space to consideration of this. I intend to argue that, from a probabilistic perspective, the existence of God in conjunction with the desire and wherewithal resulting from God's attributes, provides a confirming force on the resurrection hypothesis. Notwithstanding, there are other non-theistic arguments supporting the resurrection hypothesis. Some are naturalistic; for example, Frank Tippler's (1994) Omega Point theory and Eric Steinhart's (2014) resurrection based on digitalism. Some are spiritual but without emphasis on God; for example, Ronald Peterson (1995). Despite God being the center of my argument, I also intend to consider these non-theistic approaches to human resurrection.

Evidence is fundamental to my argument, as is variation in theistic commitment; my approach gives credence to both the atheist as well as the agnostic and theist, including the conclusion. There are a variety of ways of defining theism, agnosticism, and atheism, but for the purposes of this book, they are defined around the proposition *God exists*. The theist holding a conviction that the proposition is true, the agnostic believing it is just as false as true, and the atheist that it is false.

Despite having a personal theistic conviction, I give much value to the atheistic conviction as it represents a challenge to the theistic apologists, sharpening the quality of their arguments. I define these beliefs or convictions in terms of the probability continuum; that is, an atheist has a degree of belief in the existence of God of substantially less than 0.5 or 50% on the continuum, whereas the agnostic's degree of belief is around the 0.5 or 50% point, and the theist's substantially more. I say more about the probability continuum and these belief levels in Chapter 2.

There are three parts to my overall resurrection argument: Part I comprising a reflection on the nature of God and the validity of probability theory in analyzing evidential arguments; Part II is about the evidence that supports and refutes the existence of God; and Part III focuses on the notion of human resurrection.

Part I, Chapter 1 is about God and, if He exists, what can we say about Him; that is, given that He is relatively hidden, what defining attributes can we assign to Him? My approach in this chapter is deductive and theoretical, so conceivably acceptable to an atheist as a mere exercise in argument.

The argument for the attributes of God is a well-trodden domain and I concur with the usual analysis that God is omnipotent, omniscient, omnibenevolent, and perfectly free. Such a conclusion represents a thread that weaves through my resurrection argument and I constantly refer the reader to the conclusions of this chapter.

Chapter 2 is an introduction to the powerhouse of the inductive methodology employed throughout; that is, probability theory; more accurately subjective probability. It is a light touch approach and the non-specialist reader should be able to follow the analysis presented. For the more adventurous, I provide proofs of theorems in the endnotes and refer the reader to more detailed analysis where appropriate.

My use of the term *probability* is not intended to give authority to the analysis throughout the book, it is only a technique to *help* ensure that those arguments based upon evidence are coherent. Indeed, I argue in favor of the subjective theory of probability; that is; the idea of a degree of belief in the truth of a proposition. Despite its subjectivity, the process is substantive and its application warranted as a sound theory of probability.

An objectifying derivative of the subjective theory of probability is Donald Gillies' 1991 concept of *intersubjective probability*, whereby a group attempts to find a subjective consensus on the probabilistic force of evidence on a proposition: Chapter 2 considers this concept. A derivative of intersubjective probability is what I have termed *vicarious probability*; that is, the group attempting to reach a consensus is hypothetical and their hypothetical judgements determined by evaluators. There are both advantages and disadvantages to vicarious probability and I find in favor of the advantages, employing the procedure throughout the book.

The application of probability theory in my resurrection argument is mostly through the use of Bayes' theorem, a derivative of conditional probability developed by the Reverend Thomas Bayes. After his death, his theorem was presented to the Royal Society by Richard Price in 1763. Both the theorem and conditional probability are explained with simple examples.

Bayes' theorem does lend itself to a cumulative analysis, whereby the probability of a proposition is determined piecemeal through consideration of different supporting or refuting evidential elements one after the other. Thus, a final probability value for the proposition is determined once all the evidence is accounted for. Richard Swinburne (2004) is the doyen of this methodology applied to the existence of God.

As stated above, God is at the center of the resurrection hypothesis presented, but people hold differing views concerning God's existence and any subjective probabilistic approach to asserting God's existence should take this into account. There is a tendency in the philosophy of religion to undertake analysis from the perspective of the theist or alternatively from the perspective of the atheist. I believe they should be concurrent and include the agnostics' perspective; Bayes' theorem lends itself to such a collective approach. Hence, throughout the arguments presented I assume the views of the atheist, agnostic, and theist and no more so than in Chapters 3, 4, and 5. These chapters are concerned with the evidence in support of and abandonment of the theistic proposition *God exists*. It is a *cumulative* Bayesian approach pioneered by Richard Swinburne (2004); his approach is founded on logical probability, primarily, in his case, based upon, *inter alia*, the *a priori* principle of simplicity. However, I argue against this principle as an *a priori* criterion and favor the subjective theory of probability.

It is a mistake to judge the existence or non-existence of God on one or even a limited number of relevant evidential elements; all the elements impact on the veracity of the theistic proposition *God exists* and Bayes' theorem, in its cumulative form, lends itself to accounting for all the evidence. Indeed, it is the totality of evidence that is essential to forming a rational degree of belief in any inductive hypothesis. Thus, when considering one evidential element, no matter how large, forceful, and complex, it should always be with the view that it must dovetail with the other relevant evidential elements before a conclusion on the veracity of the hypothesis is warranted.

As stated above, God is at the center of my resurrection hypothesis; consequently, the probability of God's existence, seen through the eyes of an atheist, agnostic, and theist, is of paramount importance. To this end, the first of the three chapters of Part II; namely Chapter 3, is concerned with the evidence from divine creation; that is, the existence of the universe, the laws of nature that govern the physics of the universe, and the providential nature of the universe in terms of the potential for the emergence and development of persons and animals.

Chapter 4 engages with the most forceful evidential element in support of the abandonment of the theistic proposition; the problem of evil in terms of human and animal suffering. This is the most forceful argument in the atheists' arsenal and much ink has been spilt trying to defend what is, on face value, a most compelling refutation of the theistic proposition. I address many of the defenses, particularly the probabilistic ones and review some theodicies; arguments that attempt to explain why a God of maximal love and the actuality of horrendous evils can coexist. So negatively forceful is the problem of evil, apologists can only mitigate the probabilistic damage done to the theistic proposition and the resulting Bayesian analysis reverses any gains made in favor of the theistic proposition from the evidence presented in Chapter 3.

Having suffered a mauling from the problem of evil, the evidential analysis moves on to the evidence from the divine/human relationship (if such a relationship exists) forming the contents of Chapter 5. These traditional evidential elements are religious experience, miracles, and the *consensus gentium* argument or the argument based upon a majority of people believing in God. *Prima facie*, these arguments appear to support the theistic hypothesis. However, I argue that the first two evidential elements are bias due to their titles; that is, the titles imply the existence of God before any evidence is considered, possibly skewing any subjective probability analysis in favor of the theistic hypothesis. Thus, I argue they should have more neutral titles; *uncanny experience* in the case of religious experience and *enigmatic events* in the case of miracles.

I consider a range of arguments for and against religious experience (now termed uncanny experience) and conclude that non-supernatural interpretations are as equally valid as supernatural interpretations. However, there is one example of uncanny experience that may be different; that is, dramatic conversion experiences whereby the uncanny experience results in a sudden religious conversion. Theists and possibly agnostics would certainly favor this evidential element, but the atheist would still be skeptical. Unfortunately, the evidence in support of dramatic conversion experiences is thin and this is reflected in the Bayesian analysis that follows.

Miracles (now termed enigmatic events) have come under the critical lens of several philosophers, notably David Hume (1748) who condemned the rational acceptance of miracles based upon testimony given the laws of nature. Richard Swinburne (2004) argues otherwise and there have been some Bayesian analyses published on this issue. I consider these and the arguments of Hume, Swinburne, and others concluding that, although an agnostic and theist would probably accept the evidence of enigmatic events supporting the theistic proposition, it is unlikely that the atheist would.

The *consensus gentium* argument basically states that the evidence of the majority of people believing in God provides probabilistic support for the theistic proposition. It has ancient origins; namely, Cicero (45 BC), but is more recently associated with Loren Meierding (1998). He provides a compelling aggregation argument and an atheist, agnostic, and theist would probably accept the evidence of the majority of people believing in God supporting the theistic proposition. The Bayesian analysis is duly completed and the reality of the evidence confirmed by an Ipsos survey.

Having established, through the cumulative Bayesian analysis, the three probability values of the theistic proposition *God exists*, the reader moves on to Part III, comprising three chapters concerned with human resurrection.

Resurrection entails the preservation of identity and therefore, before proceeding with a resurrection analysis, the concept of identity needs to be established; Chapter 6 engages with this requirement. It is a complex domain of philosophical argument that employs imaginary scenarios to tease out the concept; for example, imaginary brain transplants and teleportation of *Star Trek* fame. I also included the notion of whole brain emulation in the imaginary mix, as it relates to the possibility-model of human resurrection of the next chapter.

Person-identity in the context of human resurrection is an issue that relates to the preservation of identity over time (diachronic person-identity) and across worlds (ante-mortem to post-mortem identity). The most challenging aspects of person-identity for a resurrection hypothesis are the notions of *immanent causation* and the *only x and y principle.* The former relates to the necessity of intimate causal connectivity in the preservation of identity over time and the latter to the factual exclusivity of the time-separated individuals; for example, Smith the child and Smith the adult. Their identity is exclusive to the facts and relationships between them, not that of other individuals. I address these and other matters concerning person-identity, finalizing with a definition of diachronic person-identity that I carry over to the next chapter.

In Chapter 7, I introduce a possibility-model for human resurrection; that is, a possible scenario that may fulfill the criteria of diachronic person-identity in resurrection. To test the model, I apply Scott Sturgeon's (2000) genuine possibility criteria; logical possibility, conceptual possibility, metaphysical possibility, and nomological possibility. In the process of testing the possibility-model, I adopt Lynne Baker's (2000) argument for identity through constitutionalism and include the notion of whole body and brain emulation introduced in Chapter 6. Metaphysical and nomological possibility absorb much of this chapter and I argue that my possibility-model succeeds in fulfilling the criteria of genuine possibility.

Chapter 8, the final chapter, is concerned with the probability of human resurrection and heralds the return of probability theory. Three propositions, with their own probability values, form an interrelated network; *God exists*; *Jesus died and was resurrected; dead persons will be resurrected.* An analysis of this network led to the establishment of prior probability values of the proposition *dead persons will be resurrected* from the perspective of the theist, agnostic, and atheist. Based upon these prior probability values, the final two evidential elements were evaluated; *near-death experiences* and *apparitions* which, *prima facie*, appear to probabilistically confirm the human resurrection hypothesis.

The overall argument is a chain of probabilistic analysis commencing with Chapter 3 through to Chapters 4, 5, and 8 supported by identification of the divine attributes of Chapter 1, probabilistic methodology of Chapter 2, the concept of identity of Chapter 6, and the resurrection possibility-model of Chapter 7.

I encourage the reader to participate in the probability process from the outset, suggesting that they can, if they wish, include evidential elements that I have not included. The reader's participation can extend to the final chapter and again, they may have evidential elements of their own to add in to their analysis (see Appendix 2). The result for the reader would be their subjective degree of belief in the human resurrection hypothesis and it may very well be a surprise to them.

Part I

DIVINE CHANCE

God and Probability Theory

Part 1 has two objectives: first, to lay the foundations of a fundamental conception of God, which affects the entire analysis of the book; indeed, the existence of God is at the heart of my resurrection argument. This argument is not particularly from a theological perspective, but from an attributional perspective; that is, the attributes that define God and interface with the issues that arise in my resurrection hypothesis and which act as auxiliary assumptions in determining subjective probability values. (See Appendix 1: The Auxiliary Assumption Requirement) Thus, Chapter 1 engages with the arguments related to the divine attributes, concluding with a set of necessary attributes that form the backdrop to the chapters of Parts II and III.

Chapter 2, introduces the reader to probability theory which is the analytical methodology employed throughout the book, culminating in a conclusion that provides a rational view of human resurrection in terms of subjective probability; rational that is, because the argument is governed by the axioms of probability. Thus, Chapter 2 addresses some of the issues surrounding the notion of probability and how subjective probability can be a sound methodology for determining a rational degree of belief in the truth of a proposition.

Chapter 1

Who's God? Divine Quintessence

If someone were to ask me who my next-door neighbour is, I'd probably reply "to my right is old George and his wife Ingrid and to my left are the Smiths and their three children." I'd identify them by name and if the inquirer pressed me for more information, I'd start using identifying characteristics: "you know George; at about six o'clock he always sits by the inglenook in the Rose and Crown with his Labrador and a pint of bitter." The inquirer might respond: "I know; he always wears corduroys." However, what would we say to someone if they asked, "who is God?" It's unlikely we'd say "Yahweh" or "Jehovah," and if pressed, be able to describe Him[1] like old George in his corduroys. The response is more liable to be definitional, such as "the creator of all there is" with some defining characteristics such as all-seeing and all-loving. In the case of George, I used descriptors of contingent properties that delineate him from other people. If those properties were absent, it would still be George and I would need to use different contingent descriptors. In the case of God, I must employ necessary defining characteristics of a unique being, characteristics that cannot be eliminated if the being is still to be God. Hence, to answer the question "who is God?" we can only provide the necessary and sufficient defining characteristics, which are the quintessential and essential attributes of God.

To muddy the waters even more, the idea of God is culturally variable. Some cultures see God as present in all physical things, from snails to comets, all possessing supernatural powers. Other religious cultures accept multiple gods: polytheism. The Greeks had multiple gods and goddesses with imperfect characteristics, but with specialized strengths, such as Zephyrus, god of the west wind. The ancient Egyptians had animal gods, including the baboon god Hedj-Wer. The list and variations go on, so I intend to settle upon God as understood by the three Abrahamic religions: Judaism, Christianity, and Islam.

OMNIPOTENCE

How can one start to approach defining God in terms of His quintessential attribute? I believe that one approach could be to adopt a basic attribute that is as near as universally acceptable as we can get: an axiom-like postulate. Hence, this basic attribute should be as equally acceptable to the atheist and theist alike given that definitions do not entail existence; we could define a unicorn, but it does not mean one exists. From such a basic attribute, it may be possible to derive other attributes.

ALMIGHTY

One way forward is to start with the expression '*if God exists, He would be* ....'and fill in the gap. 'Loving' and cognate terms are deemed popular,[2] but these would be too broadly applicable; a dog could be a loving pet. 'Maximally loving' would be more transcendent; notwithstanding, it is the notion of 'maximal' that makes the difference.

Many of the Greek gods were imperfect beings, but greater than humans in some respects and even awesome, such as Zeus the god of thunder. Moreover, some humans declared themselves or were declared by others as living gods like Alexander the Great. In all cases, a greatness over-and-above other humans is implied; it is even in Alexander's title. Hence, the notion of 'greatness' seems to be a contender for the basic attribute. Indeed, I remember as young teenager, watching the professional wrestler Mick MacManus on television mercilessly tearing into his opponents with a variety of semi-legal holds and blows much to the baying angst of the spectators; a *great* British wrestler. Notwithstanding, we need to distinguish clearly between human/animal contenders and God: perhaps the Anselmian conception of maximal greatness would suffice, namely "that which nothing greater can be conceived"; it certainly corresponds with the notion of *Almighty*.[3] Thus, there may be two contenders: maximal greatness and maximal love:

1. *If God exists He would be maximally great* OR
2. *If God exists He would be maximally loving*

Comparativeness

Greatness is a comparative concept and category dependent. What makes an athlete great is different from what makes a car great: different criteria apply to different categories. Athletic and automotive greatness is comparative in their respective categories; Usain Bolt is a great sprinter because he regularly beats other competitors as well as world records; an Aston Martin is a great car in certain respects compared to an Austin Allegro (apologies to Allegro devotees). There are certain qualities of Usain Bolt and an Aston Martin that make them great in the categories of sprinting and cars, but to claim that Usain Bolt is greater than an Aston Martin seems incoherent.

Given this categorisation requirement, to describe some being as the greatest, seems to be incoherent unless the category in which they are the greatest is stated. Therefore, a being can be described as the greatest if, compared to another being in *comparable categories*, that other being is found wanting. Moreover, to describe some being as *maximally great* is to claim that all reality, other than that being, possess categories comparable with that being and that, in every one of those categories, the being is the greatest.[4] There is inherent anthropomorphism in this comparative conception; that is, maximally great is the peak of the value-scale as envisaged by what humans value.[5]

Perfections

Maximal greatness implies perfection in all things, yet some things are incompatible; therefore, a being can possess perfection in some things, but not in others and consequently is not perfect in all things; and hence, cannot be maximally great. For example, the perfection of omnipresence is incompatible with the perfection of playing hide-and-seek. However, by categorizing perfection into *pure* and *mixed*, where mixed perfection entails some imperfections, such as being unable to play hide-and-seek, and pure perfection does not, then maximal greatness implies *pure perfection* and the perfections within that category are compatible. However, this could exclude omnipresence, which some apologists would balk at. An alternative is to claim that maximal perfection is that set of perfections that is the best of all possible perfections; that is, possession of all perfections is not possible, but there is a set of them that is maximal; this would save omnipresence if required.[6]

Intrinsic Maxima

It has also been argued that maximal greatness is metaphysically impossible as maximal greatness entails greatness in all things; for example, knowledge. Although some knowable things are compatible with a maximal limit (*intrinsic maxima*), such as, knowledge of all living species, others are not (*non-intrinsic maxima*); for example, knowledge of the greatest number, as there is always an additional number that can be added. Consequently, maximal greatness is incoherent.[7]

It could be argued that maximal greatness does not directly apply to an open-ended category that lacks intrinsic maxima, but does so when propositions within the category exhibit intrinsic maxima. Although an all-knowing being cannot know the greatest number, for the greatest number is impossible to know, He can know any number however, so ever defined, such as, the square root of any number. Similarly, such a being cannot possess maximal strength, as He could always be that bit stronger; however, He is stronger than any other being. Thus, a maximally great being cannot possess specific maximal qualities that are metaphysically impossible to possess. This relates to the best of all possible perfections argument above, whereby a being is maximally great if the set of qualities He possesses is the best of all possible qualities. Thus, no being, even a maximally great being, can possess qualities that are logically or metaphysically impossible to possess. Nevertheless, perhaps viewing maximal greatness as a specific comparative measure is ill conceived.

Limit Case

An alternative solution to the problem of intrinsic maxima is suggested by Barry Miller (1996). He introduces the notion of a *limit simpliciter* and a *limit case*. The former is the last measure of a progressive scale; for example, increasing the number of sides of a polygon drives the polygon towards becoming a circle, but in becoming a circle it ceases to be a polygon. This fuzzy polygon end-point is a limit simpliciter, whereas the circle is a limit case. Thus, there is a connection between these two types of limits; that is, the limit simpliciter points to the limit case. In a comparative category, wherein God is maximally great, the category comprises a set of other beings that are place-able on a scale of greatness; that is, the weakest person to the strongest person, the latter being the limit simpliciter. God, as the maximally strongest being is not on this scale, He is the limit case expressible only as a proposition rather than the last measure on the strength scale; namely, *God is maximally strong.* This echoes Anselm's notion of *that which nothing greater can be conceived.* In the case of the category of knowledge of the largest number, there is a scale of knowable numbers limited by the mental capacity of the knower, that is, a young child may know numbers up to one hundred, an adult may know numbers in the trillions. There will be a scale of number-knowledge that persons can match-up to tending towards a fuzzy limit which is the limit simpliciter. The limit simpliciter points to the divine limit case that is off the scale and only expressible as a proposition, namely, *God knows any number.*[8]

Possibility

Notwithstanding Miller's limit case argument, logical and metaphysical possibility still has application in so far as greatness, whether maximal or limited, is subject to the limitation of possibility. Examples include, coming to know a false proposition, to know the largest number, changing the past, making 2 + 2 = 5, making a square circle,[9] coming to believe He is not maximally great, and many others.[10] I would also exclude the possibility of *making* a person *freely* choose to act or for the maximally great being to commit suicide.

Possible Worlds

The limitations of possibility in divine creation can be exemplified in terms of possible worlds; namely, can God create any possible world where persons therein possess free-will? Possible world theory is associated with Leibniz who proposed that God chose to create our world from a selection of possible ones and given God's maximal greatness our world is the best of all possible worlds.[11]

Assuming free-will of persons would be prevalent in the best of all possible worlds, consider such a world that God, as a maximally great being, could create where, in that world, there is a person Mr. Itch, who can freely choose to scratch his ear. Therefore, Mr. Itch may scratch his ear or refrain from scratching his ear. Thus, there are in fact, two possible worlds, one where Mr. Itch freely scratches his ear and one where he freely chooses not to scratch his ear. God cannot cherry-pick the world to create, as that is tantamount to pre-choosing Mr. Itch's ear-scratching for him and Mr. Itch could not *freely* choose his scratching or non-scratching therein. Thus, the created world must be one offering a free-ear-scratching-opportunity and if Mr. Itch chooses to scratch therein, then God has created a world where Mr. Itch *freely* scratches his ear. Consequently, because of Mr. Itch's free-will choice, God could not have created the *freely* non-scratching possible world. The same argument holds in reverse; that is, if Mr. Itch chooses not to scratch his ear. The point of this scratching reflection is to show that, in terms of the best of all possible worlds, there are possible worlds that God, as a maximally great being, cannot create.[12]

This possibility condition does not necessarily weaken the maximal greatness of God, particularly if God instantiated the very universe where the set of metaphysical and logical impossibilities are featured. In this sense, God subjected Himself to the limitations of possibility by His creation, yet possesses the capability of reversing that creative decision.

Potentiality

The notion of greatness is frequently applied to the realization of a finite number of qualities as the above examples indicate; for example, the great athlete has actually achieved athletic success. However, maximal greatness is far from finite and largely comprises unrealized *potential*.[13] It is more applicable to consider it as latent capability as well as realized capability. The descriptor *omnipotent*, is more indicative of this latent and realized capability conjunction. Thus, although maximal greatness is a close contender for the basic attribute, the more latent capability implication of omnipotence is a better attribute descriptor. Indeed, omnipotence implies the potential for awesome supernatural deeds as well as universal maximal greatness. Hence, this third option would be:

3. *If God exists He would be omnipotent*

It is my contention that 'maximally loving' or better 'omni-benevolence' as well as other attributes can be derived from the divine attribute of omnipotence, but not vice versa. It seems that omnipotence encapsulates the notions of maximally great, maximally loving, as well as the potential for awesome supernatural action.

I do not think that omnipotence can be reduced to a simpler concept without loss of meaning. Consequently, with due consideration to metaphysical and logical possibility, I define omnipotence as:

4. *The capability to do anything that is possible to do*

CONTENDERS

Upon reflection, it seems that an evil omnipotent being could do more than a 'godly' omnipotent being because he could do all the godly things plus all the evil things too. However, I will argue that omni-benevolence is entailed by omnipotence so the evil-one cannot be omnipotent. However, to be a little pedantic, a being that could, by necessity, only scratch his ear could do everything that is possible for him to do, which isn't much.[14] An adjustment to (4) may overcome this objection:

5. *The capability to do anything that is possible to do better than any other being*

This circumnavigates the ear-scratcher example, but supposing two beings are equally matched so that one could *not* do anything better than the other;

that is to say two omnipotent beings? I argue below under Divine Freedom that this would result in neither of the two beings being omnipotent. I believe (5) settles this problem, as two potential omnipotent beings would result in no omnipotent being at all.

CONTINGENCY OR NECESSITY

In Chapter 4, I present an argument that suggests that God has withdrawn his presence from our world or even universe. Moreover, below, under the title of Omniscience, I argue that God is aware of human evil, but chooses not to intervene. These two states of affairs may suggest that omnipotence is a contingent attribute;[15] that is, the exercise of such awesome power can be switched on and off by the will of the omnipotent being. This suggests that the definition should change to:

> *The contingent capability to do anything that is possible to do better than any other being*

A proposition expressed in terms of *capability*; for example, '*an omnipotent being has the capability to...*' is not falsified by the intentional withdrawal of the force of omnipotence; *capability* suggests that it can always be reversed. In this sense '*The capability to...*' substitutes for contingency. Consequently, definition (5) still holds; moreover, (5) cannot be falsified as the loss of such capability would contradict the definition. Thus, expressed as a proposition, namely '*an omnipotent being has the capability to...*' is a necessary truth. This does not imply that God is necessarily omnipotent, just the proposition. Notwithstanding, God, if He exists, is necessarily omnipotent, so the attribute of omnipotence is impossible to eliminate from God should He exist.[16] Again, this does not imply that God necessarily exists; I address God's necessary existence below.

ANTHROPOMORPHISM

One criticism that can be levelled at this divine attribute ascription is that such attributes are human rather than divine concepts and the notion of omnipotence is anthropocentric: "Our understanding of divine power, perhaps more than any other divine attribute, shows us who *we* are." [Griffith-Dickson, 2005, 282] Thus, reflection on what attributes we admire most about persons sets the comparative archetype for the greatness of God within the category 'persons.' Within the category 'person' there are a myriad of sub-categories such as sprinting, swimming, orating, writing; omnipotence spans all the sub-categories of the category 'person.' God could, if He so wills, become human and be a champion swimmer or the saviour of mankind.

It is power, knowledge, and love that have been selected by some apologists as the most admirable attributes; hence omnipotence, omniscience, and omni-benevolence. It is my intention to weaken the anthropomorphic shackles to a small degree by arguing that it is only the first attribute that is required as the comparator in the category 'persons.' Omnipotence, albeit an anthropocentric notion, is the divine attribute from which all other attributes follow. Notwithstanding, "...if human persons are icons of the divine, then some anthropomorphism is inevitable and proper, and the label of anthropomorphism is not by itself derogatory." [Jantzen, 1984, 78]

STONED AGAIN

There have been several attacks upon the rationality of the concept of omnipotence. For example, consider the sceptic's paradox of the stone: '*could an omnipotent being make a stone so big he couldn't lift it?*'[17] If he could not either '*make it*' or '*lift it*' there is something he could not do, which is a breach of omnipotence. This hypothetical counter does not hold; consider the following:

1. There is an omnipotent being G (assumption)
2. *From (1)*: G can make a stone of any size
3. *From (1)*: G can lift a stone of any size
4. *From (1), (2) & (3)*: G can make a stone of any size and lift it
   The conclusion follows from the premises. What doesn't follow from the premises is:
5. G can make a stone so big he cannot lift it
   Indeed, if (2) and (3) are true, then (4) is true and (5) is false; that is, (5) could only be true if (3) is false. So why is there a problem? It is because (5) is usually framed as a question and the only answer is 'yes!' or 'no!' and either elicits the response '*Ah; there's something G cannot do!*.' As a proposition, (5) is either true or false and given the truth of (1)—(3) it is false. However, what does follow from (1), (2) and (3) is:
6. G *cannot* make a stone so big he cannot lift it
   Notwithstanding, (6) is a double negative and is coherent with omnipotence. It is easier to see if re-phrased:
   (6.1) *It is not the case* that G can make a stone so big he cannot lift it
   Which is the same as saying '(5) is false.'
   However, consider the following conditionals:[18]

7. *If* G cannot make a stone so big he cannot lift it *then* there is something G cannot do
8. *If* G can make a stone so big he cannot lift it *then* there is something G cannot do

   The truth value of a conditional based on the truth values of the antecedent *x* and consequent *y* are as shown in the following truth table:

Table 1.1. Truth Table for the Stone Paradox

x	y	$x \rightarrow y$
T	T	T
T	F	F
F	T	T
F	F	T

(Where → is the symbol for the conditional 'if *x* then *y*')

This table provides all the truth-values for the conditional in sentential logic.

The antecedent of the conditional 7 is expressed as 6 and 6.1 and is *true* given 1–4 above. From the truth table above, the consequent must be true for the conditional 7 to be true. Above, I argue that an omnipotent being cannot do the impossible, therefore the consequent 'there is something G cannot do' is *true*, thus, from the above truth table, the conditional 7 is *true*.

The antecedent of the conditional 8 is as 5 above and is *false* given 1–4 above. The consequent can either be true or false for the conditional 8 to be true. For the same reasons that the consequent is true in 7, so it is *true* in 8; therefore, from the above truth table, the conditional 8 is *true*.

It seems that the sceptic has won the day; however, what can be claimed by the sceptic? Not that the consequent is true because of the antecedent. The truth of the consequent is a result of an unrelated matter; that is, the impossibility of an omnipotent being to do the impossible, such as make a square circle. The conditional may hold as a logical form, but its application to this paradox does not affirm the sceptic's argument. The sceptic may respond that the consequent is true because of the antecedent and not because of this unrelated matter. However, premises 2 and 3 above, derivable from the concept of omnipotence, clearly negates such a connection. Thus, I rest the argument on the premises and conclusion of 1–4. Although the question '*could an omnipotent being make a stone so big he couldn't lift it?*' possesses linguistic meaning, it is intrinsically incoherent.[19]

THE OTHER DIVINE ATTRIBUTES

The Abrahamic God of Judaism, Christianity, and Islam is considered, by most devotees of these religions, omnipotent, omniscient, omni-benevolent, personal, totally free, spiritual, immutable, eternal, worthy of worship, and to exist necessarily, putting such a god above all possible contenders. As I have stated above, it is my intention to consider only those attributes that can be derived or refuted from consideration of the attribute of omnipotence. However, I intend to give some attention to the assertion that God is a spiritual being as this impacts on my overall argument.

Omniscience

Omniscience could be defined as '*knowledge of every possible knowable thing*.' This attribute is clearly a consequence of the basic attribute of omnipotence. How could a god be omnipotent without the necessary accompanying knowledge, be it propositional knowledge or know-how? For example, the instantiation of our universe whereby animals and plants can evolve to the point where the human animal emerges requires, *inter alia*, both the propositional knowledge of the required laws of nature and the know-how to instantiate them. Indeed, the attribute of omniscience follows from omnipotence; impossible to be all-powerful and unknowing.

It could be argued that God is not cognizant of free human mental states because such states would not be free if He were; therefore, He is not omniscient. From the above, if God is not omniscient then God cannot be omnipotent, and if God is not omnipotent then God is not God, which is illogical if the basic attribute is accepted. Consequently, given that omniscience is defined as '*knowledge of every possible knowable thing*,' it follows that it is not *possible* to know free human mental states; however, they are known by the person having them, therefore it is possible. Consequently, if it is the case that God is not cognizant of free human mental states, then there are two possibilities that save the above omnipotence/omniscience analysis from total collapse:

1. God *can know every possible knowable thing* but has *chosen* to be ignorant of free human mental states OR
2. Omniscience is defined as '*knowledge of every possible knowable thing except free human mental states*'

Option (2) changes the definition of omniscience making it more complicated than option (1); moreover, there could be even more exclusions and '*knowl-*

edge of every possible knowable thing except $exclusion_1$ *&* $exclusion_2$*...&...* $exclusion_n$' would complicate the definition even more and make it incongruous given the all-knowing element of the definition: "death from a thousand qualifications."[20] Thus, the parsimony of option (1) is the better explanation if there actually is divine ignorance of free human mental states. However, I do not accept that God is not cognizant of human mental states; we are all aware to some degree of other people's mental states through their behavior and language and their mindfulness of this awareness may influence their behavior. However, discrete or covert observation would avoid this and God's 'observation' could also be discrete and covert. Thus, the definition of omniscience remains as: *Knowledge of every possible knowable thing.*

Predetermination

A more worrying problem is divine universal awareness of future events. If God is aware of future events in the lives of humans, as for example, at next year's New Year celebrations, God knows I will get excessively merry and sing the 'Hokey Cokey,' then it is true that this bawdy event will occur way before it has occurred. Because it is pre-known, then it is *certain* to occur and I will be unable to restrain myself when the time comes. Indeed, one important criterion of the concept of knowledge, as opposed to belief, is that what is known *is true*. Hence, if it is the case that all human action is pre-known, then all future human action is predestined and free-will is a complete illusion. Worse, we cannot be held responsible for our actions or inactions because we cannot choose to so act or not to so act; Hitler and Pol Pot are as innocent as Mother Teresa and Mahatma Gandhi. This was an issue that troubled the Roman Boethius in the sixth century AD.[21]

Timeout

One possible solution, is to accept that God is outside the context of time and therefore separate from what counts as past and future events in our universe.[22] God is cognizant of our past and future in one 'outside-time-observation'[23]; thus, there are two types of knowledge:

- knowledge outside-time [24] and
- knowledge inside-time

Consider the following: an omniscient being (omni-being) perceives, outside-time, William equivocating, at inside-time t_1, over whether to sing the Hokey Cokey or not. Then, just after t_1, at inside-time t_2, William starts to sing the Hokey Cokey with no persuasion from the omni-being. Is William's free choice eliminated simply because the omni-being knew what he decided? Consider the following:

1. 'William *will* sing the Hokey Cokey at inside-time t_2'
   1.1 *At inside-time t_1*: For mere mortals, 1.1 can only be a prediction; that is, *possibly* true.
   1.2 *Outside-time*: For the omni-being, 1 is *not applicable* as there is no future; all is perceived simultaneously, that is, William is singing the Hokey Cokey at t_2. Even if it were applicable and the proposition 1 predicated with '…is true,' this would still be different from 1.1 (see 'Middle Knowledge' below). It may even be predicated with *false* as the temporal adverb 'will' has no function for an outside-time being and anything false cannot be known.[25]
2. 'William *is* singing the Hokey Cokey at inside-time t_2'
   2.1. *At inside-time t_1*: For mere mortals, this is *false* as t_2 has not arrived for William to be singing.
   2.2. *Outside-time*: For the omni-being this is *true*, as William's actions throughout his life are perceived simultaneously.

Note that the truth-values ascribable to the omni-being and mere mortals are different in both cases; there is no match, in fact with the proposition 2 there is a contradiction. In this context, what is true for outside-time is not true for inside-time and consequently, what is known outside-time is not the same as what is known inside-time. Human free-choice is inside-time and therefore, it would be foreknowledge inside-time that would be a breach of free-choice. I am not suggesting that what the omni-being perceives will not be what occurs; if He can perceive me singing the Hokey Cokey, then that is what I will choose. However, it will be a free-choice as there will be no inside-time foreknowledge of what that choice will be.

Supposing an omni-being did not exist, but instead another non-omni-being outside-time did and, moreover, this being could perceive some but not all of eternity. This being could perceive William choosing to sing or not so perceive: would William's choice not be free in the perceiving scenario and free in the non-perceiving scenario? This would be bizarre; it indicates that the mere act of perceiving eliminates William's free choice. If William knew a beady eye was upon him, this may cause him to sing or not to sing, but in this case the perception is totally secret; William's choice is uncaused by the non-omni-being.[26]

Loop Quantum Gravity

Based upon our perception of empirical change and a sense of duration, most believe time is real despite being unable to independently verify its existence;

that is, as an absolute reality rather than relational concept.[27] Its absolute reality is an assumption and has been included in the mathematics of theoretical physics. Indeed, such an ontology forms the basis of Einstein's theory of general relativity, whereby space and time are inextricably bound; a unity termed *spacetime*. If God is outside time, then according to general relativity, He must be outside space too; that is, outside the universe. However, the nascent theory of *loop quantum gravity*, that aims to reconcile the microscopic world of quantum mechanics with the macroscopic world of general relativity,[28] results (mathematically) in the elimination of time altogether, other than as an assumption at the macroscopic level. If true, God could be outside time and, simultaneously, inside the universe. Moreover, because of His omniscience, God would not be limited to any macroscopic assumptions. However, like its competitor string theory, loop quantum gravity is still only a mathematically inspired theory awaiting empirical verification.

Middle Knowledge

The central problem around the predetermination problem is that if a person can freely choose to act in a certain way, then they can also freely choose not to; they are the 'private' cause of the final result. How can an omniscient being look into their mind and come to *know* this result in advance without the issue of predetermination arising? Further, if an omniscient being cannot pre-know, then the attribute of omniscience is questionable. One possible solution is middle knowledge.[29]

The middle knowledge argument proposes three forms of divine knowledge:

1. *Natural Knowledge*: God knows His own nature and future *possibilities* resulting from both divine action and the action of free persons.
2. *Middle Knowledge*: God knows what free persons will *choose* to do.
3. *Free Knowledge*: God knows what is *actually* happening in real-time.

Middle knowledge can be acquired by prediction, which is how humans predict future events, except that God's predictive power is maximal ("super-comprehension"). Therefore, He knows the result of the free choice so that omniscience and free-will are preserved. However, this is still a case of divine prediction not divine foreknowledge, and therefore it remains just a prediction that may or may not come true.

Predictability as a method of knowledge echoes Laplace's classical theory of probability, which is how probability results from predictive failure; namely, a consequence of human ignorance of all the facts, something an omniscient being does not suffer from.[30] However, although the future can be *predicted* based upon past events, such as the sun has risen for five billion years and will arise tomorrow, which is not necessarily the case, only highly probable, albeit to a very high degree. Other predictions based upon past events are much less probable; for example, the last five tosses of this coin have come-up 'heads' therefore the next toss will come-up 'heads.' Our world is intrinsically probabilistic from the domain of particles to that of galaxies. Even if every possible past and current event is known to the finest possible degree, the very next event is still uncertain by one degree or another. This is Heisenberg's uncertainty principle of quantum mechanics and at the quantum level of particles probabilistic uncertainty is very high and at the macro level of galaxies probabilistic uncertainty is very low, but not maximally certain.[31] Thus no event in our universe can absolutely be *known* in advance other than, arguably, divine intervention and necessary truths. This is the nature of our universe and if created by God, He created it thus: God does, in fact, play dice!

Possible Worlds Revisited

An alternative middle knowledge approach based upon possible world reality,[32] is to claim that there is a possible world where a William-counterpart[33] has already got merry and *freely chosen* to sing the Hokey Cokey, which is a historical event in this possible world. Through super-comprehension, an omniscient being would be cognizant of this world and every state of affairs (including the William-counterpart getting merry) that led up to him *freely choosing* to sing. Likewise, there would be a possible world where a William-counterpart got merry and *freely chose* not to sing; an omniscient being would also be aware of this world and the states of affairs that led up to the William-counterpart *freely choosing* not to sing. Possible world counterfactual analysis claims that if the states of affairs in the actual world are more like the states of affairs in the singing-possible-world than any other possible world including the non-singing-possible-world, then the subjunctive conditional[34] '*if William gets merry then he will freely choose* (say at t_f) *to sing the Hokey Cokey at* t_2' is true.[35] Note however, that what is true is the 'freely choosing' to sing at t_f not the actual singing at t_2; that could still not occur because of a material constraint. An omniscient being would be cognizant of the best actual-to-possible world match and hence have *prior* knowledge of William's free choice in the actual world. Thus, both human free-will and omniscience are preserved, but the argument does rely upon possible world reality, divine access to human mental states and trans-world identity.[36]

The mismatch of ascribable truth-values, the non-causal nature of secret perception and trans-world identity provide warrant to the argument that outside-time omniscience answers the charge of eliminating free-will by omniscient perception.[37] One problem with this divine outside-time analysis is that it denies that God can change, as change has temporal implications. However, change is implied by divine decision making and creation and these are just what an omnipotent and omniscient being would do.[38] Perhaps God exists within His own form of time which is distinct from ours. Notwithstanding, provided God is outside our form of time, irrespective of Him being in a different form of time, then the above analysis still holds.

Just Popping Out

Just as omnipotence can be strategically withdrawn by an omnipotent being, so can omniscience. Although I argue that omnipotence and omniscience go hand in hand, omnipotence can be strategically withdrawn without affecting omniscience, but not the other way around; loss of knowledge eliminates omnipotence. To strategically withdraw omniscience is to withdraw omnipotence. However, they could be withdrawn together and in Chapter 4, I present an argument that suggests that God has strategically withdrawn from our world and possibly our universe, including divine consciousness: "Only a dynamic of the overarching power and inviolability of "quest for potential" could initiate a contraction of Divine here-and-now consciousness." [Birnbaum, 1989, 149] However, the withdrawal of inside-time omniscience does not eliminate outside-time omniscience, which, as I argue above, is quite different to inside-time omniscience. Thus, given my position on divine withdrawal from the universe, the definition of omniscience needs revision to account for such withdrawal without loss of potential:

(1) *Elected knowledge of every possible knowable thing*

Divine Freedom

Maximal-freedom is clearly a derivative of omnipotence; any extrinsic constraint upon the expression of omnipotence in action is to deny such omnipotence. However, like the restriction of impossibility on omnipotence this also applies to freedom; an omnipotent being is not free to do the illogical or impossible. Given this, the existence of two or more omnipotent beings could invade such freedom and consequently eliminate the attribute of omnipotence across all the omni-beings. Imagine two omnipotent beings and a feather; one omni-being wishes to move the feather and the other to prevent it moving at the same time. In this scenario, there would be a conflict of will resulting in the absurdity of neither being omnipotent.

I propose the following definition for maximal-freedom:

(1) *The choices of an omnipotent being to do anything that is possible to do are totally unrestricted*

The additional consequential state of affairs of maximal freedom states:

(1.1) *There is only one omnipotent, omniscient, and maximally-free being*

Omni-benevolence

The attributes of omnipotence, omniscience, and maximal-freedom provide God with the ultimate wherewithal to act in accordance with His will. Nothing, other than the impossible, can prevent Him doing what He so desires, and when done, it will correspond perfectly with those wants; perfection in action. Such action cannot be flawed in any way whatsoever; if it is, then one of these attributes would become questionable.

Suffering and Salvation

Regrettably, history shows that power, knowledge, and freedom do not necessarily result in benevolence. A person gifted with such attributes could be tyrannical and intentionally hostile to others, creating a context of suffering and fear. But can the same be said of God; can divine action still result in human and animal suffering and fear? As addressed in Chapter 4, the problem of evil indicates an absence of divine benevolence (or one or all of the other attributes). However, even the divine creation of a benevolent god may well lead to such suffering. Granular imperfections in creation may be a necessary consequence of high-level perfection. Human and animal suffering may well be necessary for the high-level perfection of human freedom and the personal development of all mankind, which is a greater-good argument. Where such granular imperfections cannot be directly ameliorated without damaging high-level perfection, alternative divine strategies can be deployed; more so where significant granular imperfections occur such as human suffering and death. Salvation and resurrection could make good this imperfection in God's creation. Notwithstanding, perhaps there is a lack of benevolence in the divine attributes; I think not. In fact, I intend to argue that the divine attributes of omnipotence, omniscience, and maximal-freedom presuppose omni-benevolence.

Divine Realism

Many moral apologists believe moral judgements to be propositional; that is, they can be true or false. For example, 'it is wrong to steal' (true); 'it is right to murder' (false). It seems reasonable to accept the truth-values of these two propositions which are indicative of objectivity in moral judgement. However, consider the proposition: 'homosexuality is wrong'; there will be considerable division on what truth-value to ascribe to this proposition. Consequently, a charge of subjectivity may be warranted, as moral judgement is just an *expression* of disapproval, or worse, merely an attempt to proselytize. However, difficulty in resolving a truth-value does not mean that moral truth-values are not applicable; moralizing may well resolve a dispute. The truth-values of natural properties are frequently disputed; for example, who has got the fastest car or which football team is best. Moral judgements can indeed be expressions of approval or disapproval, but they are not just that. They are also true or false and hours can be whiled away discussing the pros and cons of a moral issue such as 'torture by the intelligence services is wrong.' However, there is an emotional element to a moral judgement; a moral *fact* (derived from natural facts or circumstances) is a *judgement* (e.g. judged to be right or wrong) which, in turn, implies action (we should or shouldn't do this) and action presupposes a *desire* to so act or not to so act (e.g., to be bothered or not). This fact/judgement/desire conjunction presents a problem as desires are not facts or falsities (they have no truth-value), but are implicit in moral judgement and as such it is coherent for different people to exhibit a wide variety of desires in one moral judgement, which is relativity in moral judgement. It seems that moral judgement may not be as objective as first envisaged. This objection does not necessarily apply to God as His desires would not be as fickle as human desires. The assumption is that God's desires are always perfection-orientated and perfectly informed. If so, such a fixed well informed desire eliminates the possibility of variable and irrational desires across one moral judgement.[39]

Perfection in Action

Does the conjunction of the attributes of omnipotence, omniscience, and maximal-freedom presuppose a desire for perfection in divine action? Because of these divine attributes, God would know the results of any action He desires to take; conception and result would correspond perfectly: perfection in action. Notwithstanding, on first consideration it seems that an omnipotent being could perform an imperfect action by divine choice, but for what purpose? God has the wherewithal in equal measure for perfect and imperfect action: why deliberately fail when success is as simple. It is senseless, and senselessness is a weakness that would be a breach of omnipotence. In contrast, human imperfection in action is due to a weakness of some sort such as laziness, infirmity, tiredness, unconsciousness, death, indifference, lack of bodily ability, time constraints, lack of tools, lack of know-how, lack of skill, third party constraints, and so forth.

Indifference, without the intention to act imperfectly because of it, does not necessarily conflict with omnipotence, omniscience, and maximal-freedom. However, if such indifference results in a pre-known imperfection, then such indifference is blameworthy to some degree. Moreover, choice is a mental act and as such, the mental act of indifference would be a partial cause of the imperfection and therefore, an imperfect and senseless act. Thus, indifference, resulting in an imperfection, is an imperfect and senseless act and therefore a weakness and a breach of omnipotence.[40]

Moral Perfection

If they are possible to be known, God will know the *moral* truth-value related to actual and potential states of affairs and will act according to their truth or falsity. If true, God's action will be such as to uphold those states of affairs, but the opposite for moral falsity. The reason that God acts in this way is because of the attributes of omnipotence, omniscience, and maximal-freedom and, as argued above, these attributes ensure divine perfection in action and inaction.

For a person to *hope* to get things wrong or to *deliberately* act or not to act to ensure overall failure would be regarded as possessing a deranged mind. Indeed, the intentional pursuit of falsity is incongruous with human excellence and perfection. Malevolence is to pursue moral *falsity* and for God to pursue and support a state of affairs upholding a moral falsity would be an imperfection in action and breach of omnipotence. Consequently, God is a benevolent being in so far as He will uphold those states of affairs that result in moral truths and ameliorate those that result in moral falsity. Because of the force of the other maximal attributes, such benevolence is also maximal; namely, omni-benevolence, defined as follows:

(1) *The necessary maximization and amelioration of all states of affairs that result in moral truths and moral falsities respectively.*

Perfection and Free-will

If the universe was created by God, it was with the intention that it should spawn life and that would have been foreseen as a highly probable outcome (omniscience): life is an intrinsic feature of the universe and therefore was an intended part of its creation. Suffering and death of this living part of the universe are necessary for the force of evolution to take effect and for humans (and perhaps other person-lifeforms) to eventually evolve, but the legacy of suffering and death remains, augmented by horrendous human evil. *Prima facie*, this does not look like perfection in creation; however, such suffering, death and evil are, what I referred to previously as granular imperfections of divine creation, necessary for a high-level good. The emergence of humans with free-will and the capability of expressing love is an intrinsic part of a beautiful universe.

> "God allows human beings free will in order that they may be free to love Him and to show love to their neighbours. Love requires freedom; love is only possible if freedom exists. A person who is not free is unable to love." [Vardy, 1995, 113]

As I will argue in Chapter 4, it is the retraction of God from His creation that enables the emergence of free persons; there could be no granular imperfections and therefore no high-level perfection if God remained within His creation. Indeed, '*human and animal suffering is right*' is a moral falsity and the state of affairs giving rise to this falsity would be ameliorated by God if He were present, but the consequences of this alone would be to deconstruct free-will and personal development; God must retract from this universe for the sake of preserving these high-level greater goods. However, the divine force of omni-benevolence and the other divine attributes remain, albeit outside our world ensuring their fulfillment without damage to the perfection of higher-level creation. Given this, the definition of omni-benevolence is extended as follows:

(2) *The necessary maximization and amelioration of all states of affairs that result in moral truths and moral falsities respectively with the ultimate aim of realizing the greatest possible good.*

Normally, omni-benevolence is described as 'all-loving' or 'perfect-goodness'; however, I find these definitions a little hackneyed, preferring a more accurate description as above, which is one derived from the other attributes.

God the Person

Maximal greatness or omnipotence presupposes the divine quality of personhood; that is, possessing desires, consciousness, and intention. A being *incapable* of intentional action is lacking in this respect and could not be attributed with omnipotence.

A personal god is a god who possesses mental states such as conscious beliefs, desires, and intentions. Consequently, God can hear prayers and is capable of intentional action. An omnipotent being must be personal as omnipotence entails omniscience which in turn entails a being with knowledge, beliefs, and consciousness; a divine person.[41]

Worthiness of Worship

There have been and perhaps still are religions that see their god or gods possessing power, knowledge, and freedom, but such attributes do not necessarily go hand-in-hand with worthiness of worship, for such a being could be despotic and intentionally hostile to humans, creating a context of suffering and fear rather than adoration and reverence:

> "As flies to wanton boys, are we to th' gods;
> They kill us for their sport." [King Lear; Act IV; Scene I]

In ancient times, some thought that sacrificial worship was essential to repay the debt of existence and to win divine favor, such as Aztec human sacrifice. However, worthiness of worship implies that the god to whom worship is directed is intrinsically well intentioned towards the worshipers, not intrinsically hostile, requiring placatory worship.

I argue above that the attributes of omnipotence, omniscience, and maximal-freedom imply the attribute of omni-benevolence. "Indeed, it seems that a being can be *maximally* worthy of worship and moral admiration only if that being possesses all of the [these] core attributes of God." [Hoffman & Rosenkrantz, 2002, 16] *Prima facie*, it would seem that the above attributes warrant worthiness of worship, but the concept is inextricably related to human choice. Worship is an act of love and devotion to God and as such is optional and someone who has suffered great evil in the world may not feel that God is worthy of worship.

Necessary Existence

To claim existence or necessary existence as a divine attribute is a little strange; it is quite unlike predicates such as powerful or knowledgeable. To claim God exists is to state that God is not just a concept, but is real; however, in this sense God does not *possess* existence like He possesses the attribute of omniscience, He simply exists.[42] Notwithstanding, there is more to a real being than an imaginary one: there is an existential addition to the concept. Indeed, Russell argues that the ascription of 'exists' is more than just a predicate, but in addition, acts as a logical quantifier; 'X exists' states that there is one X such that 'X is [list of properties] is true.'[43]

Arguments for the necessary existence of God are termed ontological arguments[44] and are primarily associated with Anselm of Canterbury.[45] The Anselmian argument can be presented as follows:

Existence

1. *Premise*: By definition, God is the greatest possible being we can conceive; "...that than which nothing greater can be conceived" (a mental conception).
2. *Premise*: Actual existence plus a mental conception is greater than just a mental conception.
3. *From 1 & 2*:
   3.1. The God that does not exist in reality is *not* "...that than which nothing greater can be conceived"
   3.2. The God that exists in reality *is* "...that than which nothing greater can be conceived"
4. *Conclusion*: God exists.

Necessary Existence

5. *Premise*: it is greater to be something that *cannot not exist* than to be something that can *possibly* exist.
6. *Conclusion*: God necessarily exists.[46]

Existence does not necessarily make something greater in the sense of better, it can make things worse, much worse; the reality of electrocution is far worse than the thought of it: reality plus reflection is not necessarily greater than just reflection. There is an existential addition and so reality is greater in that sense, but this will not do as there is always some extra that can be added; a scale-maxima argument, which is a scale from the worst case to the best case and God is the best. I argue above, that this may not be the most appropriate way of assessing the omni-predicates and is unlikely Anselm's view given the conception *that than which nothing greater can be conceived*.[47] Non-existence would, in Anselm's view, be a flaw to this conception and contradictory.

The *conception* of God as the greatest may include His existence, but that does not entail that He *does* exist. This is Kant's problem with the ontological argument; although conceptual excellence may logically entail necessary existence, it does not ensure actual existence.[48]

From above, it seems that the greatness of God is understood because of His essential qualities; omnipotence, omniscience, and omni-benevolence (an omni-being).[49]

However, the greatness of an omni-being is only *maximal* if the omni-being exists in *all possible worlds*, as it would not be maximally great unless the being did. If something exists in all possible worlds, then its existence is *necessary.*[50] Moreover, the existence of an omni-being *is possible*. Therefore, *it is possible* that an omni-being *necessarily* exists. There is an axiom in modal logic termed S_5, which states that if a proposition is possibly necessary, it is necessary.[51] Thus, given S_5 it is necessarily true that an omni-being exists; therefore, an omni-being exists.

This Plantinga-esque ontological argument relies on the premise that it is possible that an omni-being necessarily exists which an atheist is liable to deny. However, the argument also begs the question as the conclusion is stated in the premise:

- An omni-being exists (conclusion)
- It is possible that an omni-being *necessarily* exists (premise)[52]

Consequently, although this premise may be rational, it does not establish the existence or necessary existence of God.[53] Moreover, such possible world analysis is abstract, and so not real or concrete in our normal understanding of what is real or concrete,[54] although David Lewis's vision of possible worlds is of a concrete nature, but subject to criticism.[55] If God exists in all possible worlds, He exists in our world and this cannot be changed, even by Him; this is contrary to the attribute of omnipotence. If this attribute were to be lost, possible world analysis of divine necessary existence would collapse. In fact, in Chapter 4, I argue that God has intentionally withdrawn from our world for reasons of a greater good. Thus, I do not accept that God's existence is necessary. His existence, like the existence of any concrete thing, ultimately relies upon evidence and consequently, I consider God's existence a probabilistic rather than a metaphysical issue.[56]

Divine Eternality

Factual necessity is somewhat different to logical necessity. A factual necessary being never came into being and will never go out of being. Such a being is therefore uncaused, eternal, and indestructible. Thus, if God is factually necessary, His existence is a brute fact and, in that sense, God's existence cannot be causally explained; it is a terminus of explanation.[57] Factual necessity does not entail true in all possible worlds; thus, God has the wherewithal to abstract Himself from any world, possible or factual.

If God is omnipotent and omniscient but caused, there would be a point where He did not exist and therefore would have been without power and knowledge prior to that point. This does raise the issue of the concept of time prior to the existence of our universe, assumed not to exist by cosmologists, but causation does presuppose a sequence of events that presupposes time of some sort.

God could not cause Himself for He would have had to exist prior to that cause, which is impossible. Therefore, if God is caused, something other than God caused Him to exist, that is to say, caused an omni-being to exist. Prior to that there were causal events occurring that God could not prevent or know about. It could be argued that this is a challenge to omnipotence and omniscience, thus supporting factual necessity. I prefer to use the term *divine eternality* rather than *factual necessity* so as not to confuse it with logical or metaphysical necessity.

Immutability and Incorruptibility

Given the personal nature of God, His creative agenda and the possibility acquiring new knowledge, then to claim God is unchanging is incongruous. Moreover, there is no derivation from the basic attribute of omnipotence. By contrast the attributes of omnipotence, omniscience, and omni-benevolence ensure that God is not corruptible in all senses of the word.[58]

God the Spirit

The attribute of spirituality, that is, consciousness with the ability to cause physical effects but with no spatial location, does not necessarily follow from omnipotence and the other interrelated attributes, but it is generally accepted and has implications for my overall argument.

Physical beings are located in space, whereas a spiritual being is not so located. For the three Abrahamic religions, God exists outside the physicality that has been and will be created by Him: God is a non-physical spirit that can have a causal effect on physical things. 'Soul' is a common descriptor of such a non-physical spirit: "A soul is *nonspatial substance that has mental properties*; it is a substance which is not located in space and which is conscious…God is a soul." [Hoffman. & Rosenkrantz, 2002, 39] This contrasts with the materialist view of consciousness, which is a quality or property of brain activity, a physical state of affairs. However, the above view of divine substance[59] includes a consciousness that can exist independently of the brain.

Souls and Spirits

I am extremely sceptical of the existence of a spiritual substance, if by that it is meant that the substance is non-physical with no spatial location, but possessing natural causal properties and consciousness.[60] I appreciate that this view is somewhat heterodox, but the spirit-hypothesis is an existential assumption without supporting evidence that seems to fly in the face of reason. Although much of modern particle physics and cosmology are equally bizarre, they do rest upon sound evidence and/or reason. However, do the concepts of spirit and soul necessarily presuppose non-physical conscious existence? Dualism suggests that this can be the case and perhaps this ancient dualistic perspective of human persons led to a vision of God as pure spirit.[61]

The template of the neural structure of the human brain could be described as the human soul; this is a particular materialistic view of 'soul.'[62] Perhaps divine spirit can also have a materialist corporeal nature as Hobbes suggests in his reply to Bishop Bramhall:

> "T H to his Lordship's question here: *What I leave God to be?* I answer, I leave him to be a most pure, simple, invisible spirit corporeal. By corporeal I mean a substance that has magnitude,..." [Hobbes, 1662, 313]

By contrast, Locke dismisses the idea of mind emerging from matter; "For it is as impossible to conceive that ever bare incognitive matter should produce a thinking intelligent being..." [Lock, [1632–1704], 1974, IV. X. 10] Moreover, he argues that to hold God as being corporeal is "absurd" [*Ibid*; IV. X. 16]

CONCLUSION

I began this chapter by considering how to describe God; that is, who and what is God? Unlike identifying a human by describing their contingent attributes, I advocated identifying God by first defining God's quintessential attribute, and then determining the essential attributes entailed by it. Maximal greatness was considered, which led to the adoption of omnipotence as the quintessential attribute and starting point for further analysis. From this followed a process of identifying a set of essential attributes: omniscience, omni-benevolence, freedom, and personhood. Thus, these attributes headed by omnipotence, are the defining criteria of who and what God is. The proposed definitions of these attributes were as follows:

- Omnipotence: *The capability to do anything that is possible to do better than any other being.*
- Omniscience: *Elected knowledge of every possible knowable thing.*
- Omni-benevolence: *The necessary maximization and amelioration of all states of affairs that result in moral truths and moral falsities respectively with the ultimate aim of realizing the greatest possible good.*
- Divine Freedom: *The choices of an omnipotent being to do anything that is possible to do are totally unrestricted.*

In addition to the above core attributes, I accepted the following essential divine attributes: uniqueness, worthiness of worship, personhood, divine eternality and incorruptibility.

In arguing this position, many defensive encounters were fought and other non-essential attributes considered. One controversial divine descriptor introduced was the possible corporeal rather than spiritual nature of God.

To the reader, the above analysis may seem secondary to my overall objective; that is, considering the probability of human resurrection. However, my probabilistic analysis follows a path that includes the probability of the existence of God and the attributes of God, if He exists, are essential to that probabilistic analysis. Consequently, I have attempted to be reasonably thorough in exposing my thoughts on this matter.

NOTES

1. I employ the masculine third person pronoun to refer to God with an upper case first letter: Him; He; His. The use of the masculine gender is not intended to imply anything whatsoever, but only to follow a historical trend. Indeed, predicating God with a gender at all is questionable.
2. Baylor Religion Survey, 2005—The Association of Religion Data Archives.
3. Anselm of Canterbury (1077); *Proslogion* II.
4. Wainwright W. J. (1999); p. 6f.
5. Miller B. (1996); p. 6 and herein Ch. 1; *Anthropomorphism*.
6. Wainwright W. J. (1999); p. 7f.
7. Hoffman J. & Rosenkrantz G. S. (2002); p. 17.
8. Miller B. (1996); pp. 7–10.
9. Descartes took the rather singular view that God can do the illogical, e.g. make a square circle. This denial of logical impossibility is termed 'volunterism': Cottingham J, Stoothoff R., Dugald Murdoch D., and Kenny (Trans.) (1985/1991)—*The Philosophical Writings of Descartes*—Cambridge: Cambridge University Press; 2:294 (Sixth Replies) and 3:23–26 (letters to Mersenne).
10. Swinburne R. (1997); Ch. 9 for a more thorough analysis of this point.
11. Blumenfield D. (1995).
12. Plantinga termed this 'Leibniz's Lapse': Plantinga A. (1982); pp. 180–184.
13. By potential, I do not mean possible or conceivable, but rather potency or capacity to realise a result: Aquinas T. [1274] (1991); 25.1.
14. Plantinga A. (1990); p. 170; also, La Croix R. (1977)—*The impossibility of defining 'omnipotence'*—*Philosophical Studies* 32 (2); pp. 181–190.
15. By *contingent* I mean that a proposition expressing the attribute can be true at one moment or place and false at another moment or place. This is opposed to a *necessary* proposition which is true always and in all places. Swinburne R. (1977); Part II for a more extended analysis of divine contingency.

16. This difference is of the form *de dicto* necessity for the proposition and *de re* necessity for an attribute of something. Hoffman J. & Rosenkrantz G. S. (2002); p. 168 for an additional view on omnipotence as a necessary attribute of God, should He exist.

17. Savage C. W (1976)—*The Paradox of the Stone*—Philosophical Review pp. 74–79.

18. A conditional is expressed in English as 'if…then…'.

19. For additional defences on this matter: Swinburne R. (1997); Ch. 9; Mavrodes G. I. (1963)—*Some Puzzles Concerning Omnipotence;* Philosophical review 73; pp. 221–223; Frankfurt H. (1964)—*The Logic of Omnipotence*—Philosophical review 73; pp. 262–263.

20. This is a famous quote from Anthony Flew in Flew A. (1950); p.1.

21. Boethius A. M. S. [C6th] (1999); Ch. XLI; 2.

22. *Ibid*; Ch. XLII.

23. Aquinas T. [1274] (1991); Ch.2; 14.13.

24. I have not employed the expression 'eternal knowledge' which has temporal implications. By outside-time I mean a total disengagement from time.

25. Gerard J. & Hughes S. J. (1998), p. 87.

26. Vardy P. (1995); p. 124.

27. We have a sense of dynamic time, namely event *e* has a past, present, and future timeline, and a sense of static time, such as *e* occurred on a certain date.

28. The problem being that quantum mechanics does not explain gravity as general relativity does; they are incompatible. Thus, hope of a unified theory through quantum mechanics was thwarted until the theory of loop quantum gravity emerged. Rovelli C. (2016); Part 3 for a clear non-mathematical description of loop quantum gravity.

29. The middle knowledge argument was first expounded by Luis de Molina (1535–1600).

30. David F. N. (1962).

31. Hawking S. & Mlodinow L. (2010); pp. 70–72; also Greene B. (2011); Ch. 8 for a short and accessible account of Heisenberg's uncertainty principle.

32. Lewis D. (1986); pp. 84–91 & Plantinga A. (1982); Ch. 4.

33. By 'counterpart' I mean an exact replica of William, but not the William of the actual world; Plantinga A. (1982); p. 102ff.

34. A subjunctive (or counterfactual) conditional is of the form '*if p then possibly q*'.

35. Plantinga A. (1982); p. 174ff.

36. Plantinga A. (1982) for a fuller analysis of counterpart theory, trans-world identity related to human fee-will. Adams R. M. (1990) for a counterargument against the middle knowledge argument, including the trans-world identity approach.

37. The outside-time argument is related to Birnbaum's theodicy introduced in Chapter 4, i.e., divine strategic withdrawal from our universe and the implications for the problem of evil thereto: Birnbaum D. (1989); p. 136ff.

38. Jantzen G. M. (1984); Ch. 4.

39. Swinburne (1977) also adopts an objective moral-fact position, but a naturalist one; i.e., natural (non-moral) facts can include moral properties that possess truth-value. (Ethical naturalism) Thus, human behavior can include moral properties and if the same behavior is replicated with different people at a different time, then that too includes the same moral properties possessing the same truth-value. [Swinburne R. (1977); pp.188-216] Likewise, Ethical Intuitionism is also an objective position where moral facts are directly or intuitively comprehended. [Dancy J. (2003); pp. 411–419] However, not all ethical theories accept objectivity, e.g., subjectivism and relativism. Singer P. (2003) for a good introduction to ethics.

40. It could be argued that culpability only applies to caused imperfections not to permitted imperfections. Duty of care in English tort law applies to both actions (jumping traffic lights) and omissions (forgetting to switch on car lights), but where they are the direct cause of harm. This is different to non-intervention, i.e., standing by when intervention can help prevent harm. Although there is normally no liability in law through non-intervention, there is if the non-intervener has the responsibility of intervening, e.g., a doctor. Notwithstanding, most would deem it immoral to witness a fire destroying a house and not attempt to call the fire brigade. I suggest both caused and permitted omissions can occasion moral culpability.

41. In Chapter 6, I define person-identity as *a unique being with a complex mental/noetic system and physical body*. The attribute of a physical body does not fit with a spiritual person which, if the case, suggests that either the notion of a divine person is metaphorical or the definition is false. However, pantheism and panentheism take on a physical perspective of divine personhood.

42. Kant I. [1781] (1929); Ch.,3, Sec.,4.

43. Russell B. (1956); pp. 41–56.

44. Ontological arguments are about the nature of being or existence.

45. Anselm of Canterbury (1077)—*Monologion* ; *ibid* (1078)—*Proslogion*—Chts. 2 & 3.

46. Griffith-Dickson G. (2005); p. 226f.

47. *Limit case*; Ch. 1.

48. Kant I.$_{1781}$ (1929); B.621, B.634.

49. Plantinga refers to the possession of these attributes as possessing 'excellence' if restricted to one possible world and possessing 'greatness' if applicable to all other worlds: Plantinga A. (1996); p. 107f.

50. A definition of necessary existence could be that a proposition expressing such existence is necessarily true if it is true in all possible worlds; Plantinga A. (1982); Ch. IV, 8.

51. The modal axiom S_5 is given by the notation $\Diamond\Box p = \Box p$ where $\Diamond$ means 'it is possible that…' and $\Box$ means 'it is necessary that…' Thus, S_5 states 'it is possible that necessary *p* is identical to necessarily *p*. Plantinga A. (1982); p. 52.

52. Sobel J. H. (2004); p. 96.

53. Plantinga A. (1996); p.112 and Sobel J. H. (2004); p. 96f.

54. Hoffman J. & Rozenkrantz G. S. (2002); pp. 24–26.

55. *Ibid*; pp. 82–88.

56. Ontological arguments for the existence of God are numerous from Anselm to the present day and I would direct the interested reader on this subject to Sobel, J. H. (2004); pp. 29–114.

57. Swinburne R. (2004); p. 96.

58. Hoffman J. & Rosenkrantz G. S. (2002); Ch.5; 5.3.

59. The notion of 'substance' is primarily a philosophical term referring to that which owns properties or attributes, e.g., this *apple* is green, round, shiny, hard, tasty, living.... Thus, strip away (in the mind's eye) all the properties of the apple and what remains is the substance of apple-ness; the enduring remnant. After a few moments reflection, you will realize this is quite complex philosophical issue. Wiggins D. (1998); pp. 214–249 for a good introduction to the topic.

60. However, quantum mechanics postulates a quantum field comprising particles with no spatial extension, e.g., photons.

61. Jantzen G. M. (1984); p. 2f.

62. Such a template is not a substance that occupies space, but a variable neurological layout that, *inter alia*, generates consciousness and that could be recorded ante-mortem and emulated post-mortem.

Chapter 2

A Sporting Chance

Probability

In the domain of probability theory there are tools we can use to help determine the probable truth of an existential proposition such as *there are fairies at the bottom of the garden* or in this case *God exists*; such tools harness evidence in a sound and rational way to confirm or disconfirm the proposition to a degree. My intention in this chapter is to take the reader through a basic introduction to probability theory and how I intend to employ it in my overall argument. I provide references for further reading should the more enthusiastic reader want to follow-up on the theory and principles presented.[1]

THE THEORIES OF PROBABILITY

The common view of the notion of probability is statistical probability or more accurately objective probability, whereby a significant amount of data is collected and a conclusion extrapolated from it; for example, there are 55 tabby cats in one square mile of the borough and given the borough sits in ten square miles there is, conceivably, close to 550 tabby cats in the borough. On the other hand, it is not uncommon to see commuters scurrying along to catch their train which they believe will be pulling in to the platform in a trice; their behavior being motivated by memory rather than a calculation based upon the frequency of a long series of train-events. This is also a form of probability; they believe the same train-event will repeat itself as it has on all the previous occasions; they would put money on it. The tabby cat example is a form of objective probability, more precisely frequency theory.[2] The commuter example is a form of epistemic probability,[3] more precisely a degree of belief type of probability.[4]

The probability of the truth of the proposition *God exists* falls in to the epistemic camp and there are three theories of probability in this domain: the classical theory, the logical theory, and the subjective theory.[5]

The Classical Theory of Probability

Pierre-Simon Laplace (1814) is commonly associated with the classical theory, but it can be traced back to Pascal and Fermat (early 17th-century French mathematicians).[6] Laplace argues that the universe is purely deterministic (all effects have a unique cause) and an illusion of probability emerges from man's lack of ability to fully comprehend the minutiae of cause and effect. Thus, according to Laplace, probability can only be applied to equally likely events such as throwing a fair die.[7] The most obvious refutation of this theory is the indeterminate nature of quantum mechanics (mathematics of the indeterminate behavior of sub-atomic particles).[8]

The Logical Theory of Probability

The logical theory is somewhat like entailment in sentential logic where a conclusion necessarily follows from the premises; however, in this case the premise partially entails the conclusion.[9] More simply put, if our belief in a proposition p partially entails a proposition q then we should believe q to the same degree. The logical theory is problematic as it relies upon the principle of indifference; a principle that leads to several paradoxes.[10] The philosopher Richard Swinburne (2004) employs a probabilistic argument based upon a form of the logical theory; that is, applying the notion of simplicity as an *a priori* [11] 'yardstick' in probabilistic judgement.[12]

The Subjective Theory of Probability

The subjective theory is, as the name suggests, personal; that is, the personal credence someone gives to the chances of an event occurring expressed as a proposition; for example, 'the train will be on time.' Their justification for such a belief stems from how they view the support some evidence gives to their belief; such as how many times the train has been late before. More formally, such support or lack of it is expressed as a *rational* degree of belief in the hypothesis or proposition given the evidence. *Prima facie*, this theory seems a little vague compared to its more objective cousins. However, as the scurrying commuters will testify, it is frequently used by us all. Moreover, it is not as vague as it seems; it can be given objectivity with supportive procedure, such as asking other rushing commuters "*are we too late?*" an intersubjective procedure. Further, it can be coherent with the axioms of probability (see below). Subjective probability must not gain credence from its title; that is, be directly associated with the objectivity of statistics, for example. It is simply a degree of belief warranted by its relationship to the axioms of probability.

THE AXIOMS OF PROBABILITY

There are alternative axioms of probability, but the ones widely accepted are the Kolmogorov axioms[13]. Given that the probability functions throughout this book relate to the truth or falsity of propositions, I express these axioms in their propositional form as follows:

Let p be a proposition; for example, *God exists*, then:

(1) *Axiom 1*:

The first axiom states that the probability value of p is a non-negative real number; zero or above; that is, $\Pr(p) \geq 0$

Although the axiom does not specify how large this number can be, it eliminates the possibility of negative numbers with zero being the smallest probability value; that is, in terms of propositional probability, certainly false.

(2) *Axiom 2*:

The second axiom sets the maximum probability value that can be ascribed; if proposition p is certainly true, then $\Pr(p) = 1$

$0 \leq \Pr(p) \leq 1$ follows from axioms 1 and 2

(3) *Axiom 3*:

The third axiom, known as the addition law, states that if propositions p and q are mutually inconsistent (only one can be true), then $\Pr(p \vee q) = \Pr(p) + \Pr(q)$ where v symbolizes a disjunction meaning 'or.'

From axiom 3, it can be deduced that any number of propositions can fulfill the addition law provided they are mutually inconsistent.

The probability of the proposition *God exists or does not exist* is certain; that is, a probability of 1. This is because either God exists or not, but God cannot both exist and not exist; mutual inconsistency clearly applies. Thus, from axiom 3 we have $\Pr(p \vee \neg p) = \Pr(p) + \Pr(\neg p) = 1$ where p is the proposition *God exists* and $\neg$ symbolizes negation meaning 'not.'

THEOREM OF TOTAL PROBABILITY

A useful theorem is that of total probability that I employ in later chapters. The theorem can be presented thus:

$$\Pr(p) = \Pr(p|q)\Pr(q) + \Pr(p|\neg q)\Pr(\neg q)$$

The theorem need not be limited to two propositions, but can be extended to account for any number such as: $\Pr(p) = \Pr(p|(q \,\&\, r))\Pr(q \,\&\, r) + \Pr(p|\neg(q \,\&\, r))\Pr(\neg(q \,\&\, r))$ where $\Pr(q \,\&\, r) = \Pr(q)\Pr(r)$ and $\Pr\neg(q \,\&\, r) = \Pr(\neg q)\Pr(\neg r)$ if propositions q and r are independent; that is, one does not affect the probability of the other.

COHERENCE

Subjective probability would have substantial credence if it could be mathematically related to these axioms. Donald Gillies (2003) demonstrates that it is indeed related through the notion of Coherence; a central facet of the subjective theory. He demonstrates that Coherence is both necessary and sufficient for the axioms of probability.[14]

Coherence in this context means much more than 'consistency'; the concept is derived from betting as an action, a notion proposed almost simultaneously by Frank Ramsey (1926) and Bruno De Finette (1930).[15] Their theories were based upon betting quotients (reflecting a degree of belief) providing the all-important numerical measurability of number between 0 and 1. Coherence is defined as avoiding a Dutch-book bet.[16]

Dutch-book

A traditional context for betting is horse racing, with a punter and a bookie. The bookie sets the odds and the punter bets against the bookie; that is, the punter believes one of the horses he chooses will win, but the bookie is betting against the punter; that is, the bookie believes, to varying degrees, that those horses will not win. The bookie's beliefs are reflected in the odds he has chosen. The following expressions apply:

- *Odds*: these are numerical and usually against the horse winning. There are two numbers expressing the odds; for example, 3 to 1 against. Here the bookie is claiming that a horse is three times more liable to lose than win. The punter, if he places a bet is expressing a belief, the degree of which is reflected by the sum of money laid down, that the horse is still liable to win.
- *Betting quotient*: If the horse and hence the punter does win, the bookie will pay the punter in accordance with the odds quoted, in the above case three times the amount of the bet placed. However, the sum bet is returned to the punter and this is included in the betting quotient. Thus the betting quotient, in the case of the 3 to 1 against odds is expressed as follows: $1/(3 + 1)$.
- *Probability*: the probability of the horse winning in the bookie's lights is the numerical value of the betting quotient; that is, 0.25.
- *Dutch-Book*: this is where the odds are set by the bookie in such way that despite one of the horses picked to win by the punter does win, the bookie still takes more money off the punter than is returned to him.[17]

CONDITIONAL PROBABILITY

Conditional probability is the probability of an event occurring given that another related event has occurred; for example, the probability of a cheetah sprinting given an antelope is meandering by. Herein, I am concerned with the probability of propositions rather than events; for example, 'the cheetah is sprinting,' although conditional probability applies to both.

There is also an important definition of conditional probability:[18]

Let *p* state *God exists* and *e* represent some relevant evidence; for example, evidence of evolutionary theory; then:

$$\textit{Conditional probability}\text{: } \Pr(p|e) = \frac{\Pr(p \,\&\, e)}{\Pr(e)} \text{ provided } \Pr(e) \neq 0$$

$\Pr(p|e)$ this means the probability of *p* given or assuming the truth of *e*. Conditional probability is indicated by the vertical line | in the function. It is defined on the right-hand side of the equals sign; that is, the probability of *p* and *e* together divided by the probability of *e*.

Donald Gillies (2003) also demonstrates that Coherence is necessary and sufficient for the definition of conditional probability.[19] This gives further credence to the subjective theory over the logical theory, as in the logical theory the axioms are justified by an appeal to intuition, whereas in the subjective theory the axioms and the definition of conditional probability can be proved from the condition of Coherence. Moreover, by not appealing to the principle of indifference, the paradoxes referred to above are resolved.

Alvin Plantinga (1988) questions the rationality of subjective probability. He argues that, "The suggestion that coherence is *sufficient* for rationality is, of course, at best monumentally dubious; a person's credence function can be coherent even if his beliefs never so much as change in response to changes in his experience." [Plantinga, 1988, 80] On this point, Plantinga is wrong; subjective probability is based upon Coherence, which is avoiding a Dutch-Book by adhering to the axioms of probability. A change in experience implies additional evidence and if it is not accounted for then Coherence fails to apply. If subjective probability is defined as a degree of belief by a person, then $\Pr(q|e)$ is that person's degree of belief in q given the evidence e. If there is evidence e and no degree of belief is formed on it, then the person has not (metaphorically) engaged in a bet and therefore, his credence function is not Coherent in so far as there is nothing to cohere to. It is rather like a punter about to place a bet with a bookie having read the form of the horse, changing his mind and walking away: no bet, no Coherence.

Plantinga does raise a further issue of rationality and probability in terms of the mental capacity of, in betting terminology, the punter. The rationality of a bet by someone does depend upon "*...someone whose noetic faculties were functioning properly.*" [Plantinga, 1988, 84] Thus, a misinterpretation of the evidence due to noetic incapability could result in a Coherent yet irrational bet. Thus, from this point of view Coherence is not sufficient for rationality, but nevertheless, it is necessary. To some degree, one solution to poor noetic hygiene is the process of intersubjective probability.

INTERSUBJECTIVE PROBABILITY

Staying within the constraints of the axioms is clearly the most important procedure in the application of subjective probability, but there are other options for the analyst. Donald Gillies' (1991) notion of intersubjective probability is one of them; that is, reaching a consensus in a diverse group or *syndicate* rather than an individual's degree of belief.[20]

The scurrying commuters, referred to above, could be in a group with some believing they are late, some on time, and some early. Assuming they are all fit and able to discuss the matter on the hoof, they could reach a consensus and adjust their pace accordingly. Interestingly, Gillies D. (2003) also demonstrates that intersubjective probability, provided it is Coherent as discussed above, satisfies the axioms of probability.[21]

The process of intersubjective analysis requires additional conditions to maximize its effectiveness. Gillies D. (2003) suggests two conditions:

1. *Common Interest*: this refers to a common purpose in reaching a consensus. Disagreement is to be expected, but this can have the effect of intensifying the discussion and, given the common interest in reaching consensus, a more reflective outcome can result. A consensus regarding a probability value is not an average of the syndicate values, but an agreement by all members of the syndicate for a single value.
2. *Flow of Information*: relevant data such as available evidence and third party analysis, should be made available to all members of the syndicate. An interesting point relates to Bayes theorem (see below) where background knowledge (k) forms part of each function of the theorem. Pooled background knowledge from the syndicate would probably be more extensive than that of an individual; this would add to the collective lens placed upon the issue under consideration. Moreover, third party consultation of an authority on the issue being considered by the syndicate would also add to the pool of available knowledge.[22]

Although these two conditions are essential, they may not be sufficient to optimize potential success in the intersubjective process. Indeed, setting objectives, information gathering and sharing, discussion and analysis of the information, predictions, and alternatives require more activity than just ensuring a common interest and a flow of information. The process must secure a full and managed discursive interaction between all members of the syndicate in a context of lateral thinking and free expression, where all views are given due consideration. Thus, I add the following conditions:

3. *Context of Reasoning*: reasoned consensus is more liable to create a successful syndicate than one based upon intimidation, indoctrination, or even intuition. A conclusion that has been formed from reason is *more liable* to be successful, simply because due consideration is given to the confirming force of the evidence.
4. *Breadth of Evidence*: Reasoning implies knowledge and understanding of that being reasoned and individual members of, or consultants to the syndicate may possess particular and relevant knowledge and understanding not possessed by other members. A wise syndicate should harness such diverse resources.
5. *Moderation Process*: given the desire for a sound reasoning process and the potential for conflict within the syndicate, it would be well to include a moderator to introduce ideas, stimulate discussion, and control conflict and summaries argument.

BACKGROUND KNOWLEDGE

Taking into account all relevant knowledge in a subjective or intersubjective probability analysis will add further objectivity to the process. Indeed, consideration of all the relevant evidence, whether directly or indirectly as background knowledge is essential to the rational process; a principle of total evidence: "In the application of inductive logic [23] to a given knowledge situation, the total evidence available must be taken as a basis for determining the degree of confirmation." [Carnap, 1950, 211]

Clearly, in many cases it is impossible to gather total evidence. The detective's job would be so much simpler if all the evidence was available for all her cases. However, the principle holds and every effort should be made to account for all the relevant evidence in a probabilistic analysis.

BAYES' THEOREM

Bayes' theorem[24] can be derived from the definition of conditional probability.[25] The theorem is expressed as follows:

$$\textit{Bayes' theorem}\text{: } \Pr(p|e\ \&\ k) = \frac{\Pr(e|p\ \&\ k)\Pr(p|k)}{\Pr(e|k)}$$

p is normally expressed as *h* for hypothesis, but is effectively the same. *k* refers to background knowledge; that is, not the evidence *e*; indeed, all the relevant knowledge that could impact upon the analysis. The vertical line | is the same here as for conditional probability; for example, Pr(*p*|*e* & *k*) the probability of *p given or on* the evidence *e* (assumed true) and background knowledge *k*. The various functions of Bayes' theorem have descriptive titles as follows:

- Pr(*p*|*e* & *k*): posterior probability

This function is what interests us the most if we are looking for the probability of *p* given the evidence *e* and background knowledge *k*. It is determined from the other functions in the theorem. It is worth noting that the posterior probability has a counterpart; Pr(¬*p*|*e* & *k*). This is the probability of *p* not being true given the evidence *e* and background knowledge *k*. It is also worth noting that Pr(¬*p*|*e* & k). = 1-Pr(*p*|*e* & k) (refer back to axiom 2 above) given that they are exclusive and exhaustive; this means that $\Pr(p|e \,\&\, k) + \Pr(\neg p|e \,\&\, k) = 1$ (refer back to axiom 3 above).

- $\Pr(e|p \,\&\, k)$: likelihood

This is a most interesting function of the theorem; it is the probability of the evidence assuming the truth of the proposition and background knowledge; that is, how likely is the evidence to be found if the proposition is true; it is what Sherlock Holmes, Conan Doyle's detective, refers to as "reasoning backwards."

- $\Pr(p|k)$: prior probability

This is the probability of p on background knowledge k before (or *prior* to) reflection on the probabilistic force of the evidence e. In the case of the proposition that concerns us herein and given the subjective theory of probability, it is that initial degree of belief in the existence of God on background knowledge alone such as knowledge of science, liturgy, upbringing, and news.

The prior probability function is where most criticism is directed when considering the theistic proposition *God exists*. The atheist will accuse the theist of irrationality and the theist will accuse the atheist of myopia and both believe they are right. Irrespective of such polemic interaction, the prior probability value in subjective probability is a person's degree of belief in, in this case, the theistic proposition however formed. Such belief can be formed from a whole host of reasons; for example, total faith in science to total faith in a particular religion. Extreme positions on the probability continuum would be indicative of a fundamentalist view, be it extreme atheism or extreme theism. Such entrenched conviction is liable to lead to a rejection of confirming and disconfirming evidence which should be accounted for in, for example, the formation of an intersubjective group seeking consensus on confirming or disconfirming evidence on the theistic proposition.

- $\Pr(e|k)$: expectedness

This is the prior probability of the evidence on background knowledge alone (or what evidence is *expected* to be found without consideration of a proposition or hypothesis). Interestingly, if it is unexpected and unusual, then $\Pr(e|k)$ would be low. Look at Bayes' theorem above, $\Pr(e|k)$ is the denominator, therefore if it is low it raises the posterior probability $\Pr(p|e \,\&\, k)$.

Descriptively, Bayes' theorem can be expressed as follows:

$$\text{Posterior probability} = \frac{\text{Likelihood} \times \text{Prior probability}}{\text{Expectedness}}$$

The Comparative Form of Bayes' Theorem [26]

Bayes' theorem is frequently employed to compare a hypothesis with its negation on the same evidence and by some rearranging of the theorem we have:

$$\frac{\Pr(p|e\ \&\ k)}{\Pr(\neg p|e\ \&\ k)} = \frac{\Pr(p|k)}{\Pr(\neg p|k)} \times \frac{\Pr(e|p\ \&\ k)}{\Pr(e|\neg p\ \&\ k)}$$

Objections to the Theorem

The application of Bayes theorem has not been without its dissenters, but criticisms of the theorem have elicited good counter arguments and Bayes' theorem is still alive, robust, and kicking and is applied in a range of domains, including philosophy of religion, business, engineering, medicine, science, politics, computer science, and law to name but a few. Should the reader wish to follow-up on these critical arguments and responses, a good starting point would be John Earman (1996); however, such consideration is beyond the scope of this book.

Confirmation and Disconfirmation

I have used these expressions earlier, but what do they actually mean? They are terms that apply to conditional probability; for example, in $\Pr(p|e\ \&\ k)$ p is *confirmed* by $e\ \&\ k$ if $\Pr(p|e\ \&\ k)$ is greater than $\Pr(p|k)$; that is, the posterior probability is raised above the prior probability because of evidence e. If the reverse occurs; namely, if the posterior probability is reduced below the prior probability because of evidence e, then *e disconfirms p*. For example, take the proposition 'Sherlock Holmes died at the Reichenbach Falls' (q); Watson's prior probability of the truth of q was high. Indeed, he lamented "Any attempt at recovering the bodies was absolutely hopeless..."[27] If Holmes' cloak had been discovered in the swirling waters, such a discovery would have raised Watson's prior probability of the truth of q. This increase in the prior probability of q by evidence is termed 'confirmation.' However, quite the opposite occurred: sometime later Holmes appeared in front Watson evoking the exclamation "Holmes is it really you?"[28] This sudden and incredible reappearance dramatically lowered Watson's prior probability of q. This decreasing of the prior probability of q by evidence is termed 'disconfirmation.' In these

Holmesian examples, I dropped background knowledge *k* for simplicity, but of course, it plays an important role.

Likelihoodism

An alternative to Bayes' theorem is that of likelihoodism; that is, employing comparative likelihood values alone without reference to the prior probability and expectedness functions of Bayes' theorem with their inherent assessment difficulties. With likelihoodism, the posterior function of Bayes' theorem cannot be determined.

With different hypotheses, likelihood values are indicative of how the evidence supports the different hypotheses; such as $\Pr(e|h_1) > \Pr(e|h_2)$. However, unlike Bayesianism, likelihoodism does not directly engage with belief change. Moreover, likelihoodism does not render itself to a cumulative process; that is, determining how different evidential elements can incrementally change a degree of belief in a hypothesis. Notwithstanding, likelihood ratios when used with Bayes' theorem, can be both cumulative and engage with belief change.[29]

Likelihood Ratios

Taking *p* to be the proposition *God exists*, then from the axioms of probability $\Pr(p) + \Pr(\neg p) = 1$ (exclusive and exhaustive). For example, if the probability of the existence of God is 0.25 (a possible atheist's position) then the probability of God not existing in the atheist's lights is 0.75 (fairly high). An agnostic's position is $\Pr(p) + \Pr(\neg p) = 0.5 + 0.5 = 1$ and a possible theist's position could be $\Pr(p) + \Pr(\neg p) = 0.75 + 0.25 = 1$. This dualistic vision of the probability of the proposition *p* provides a comparative opportunity in Bayesianism in the form of likelihood ratios. Stephen Unwin (2003) presents a comparative version of Bayes' theorem:[30]

$$\Pr(p|\text{e} \ \& \ \text{k}) = \frac{\lambda \times \Pr(p|k)}{[\lambda \times \Pr(p|k)] + [1 - \Pr(p|k)]}$$

$$\text{Where } \lambda = \frac{\Pr(e|p \ \& \ k)}{\Pr(e|\neg p \ \& \ k)} \text{ (the likelihood ratio } (\lambda))$$

A numerical example may help clarify the use of the likelihood ratio:
Let:

e mean there is an antelope meandering close by
c mean the cheetah will begin to sprint (I'm using *c* instead of *p*)

$Pr(c|k) = 0.2$ (cheetahs generally laze about)
If $Pr(e|c \,\&\, k) = 0.5$ and $Pr(e|\neg c \,\&\, k) = 0.1$ then $\lambda = 5$

Then using the adjusted form of Bayes' theorem above we have:

$$Pr(c|e \,\&\, k) = \frac{5 \times 0.2}{(5 \times 0.2) + (1 - 0.2)} = 0.55$$

Thus, it is just more probable than not that the cheetah will to break into a sprint.

Bayesianism can be a comparative tool; that is, comparing different hypotheses or propositions and in this chapter I am comparing two propositions that are exclusive and exhaustive: *God exists* and *God does not exist* (both cannot, but one must prevail) and the above contortion of the theorem presents a very useful tool in this process.

For the subjective Bayesian, likelihood ratios are advantageous, for it is a single subjective judgement; a comparison between one likelihood value and another. It removes the necessity to provide two specific subjective values for the likelihoods $Pr(e|p \,\&\, k)$ and $Pr(e|\neg p \,\&\, k)$. With such ratios, over or under evaluating likelihoods become less crucial as it is the comparative result that is paramount; for example, 0.4/0.2 = 2 and 0.2/0.1 = 2. The expectedness function Pr(e|k) is also becomes redundant with the likelihood ratio form of Bayes' theorem.

Auxiliary Assumptions

When determining the value of a likelihood, much depends upon assumptions related to the hypothesis. For example, take the likelihood $Pr(e|h \,\&\, k)$, where *e* represents the evidential proposition *Smith will be late for her appointment*, *h* the hypothesis *the train is late* and *k* relevant background knowledge. To allocate a subjective value to $Pr(e|h \,\&\, k)$; that is, to form a degree of belief in the truth of *e*, there must be a probabilistic relationship between *e* and (*h* & *k*), namely the relationship must be well-defined. However, in this case it is not unless assumptions are included in *h* & *k*; for example, a_1: *Smith is waiting for the train* and a_2: *if the train is late at pick-up then it may be late at its destination.* Such propositions are assumed and can be included in the function $Pr(e|h \,\&\, a_1 \,\&\, a_2 \,\&\, k)$ or in background knowledge *k*. Such assumptions are termed *auxiliary assumptions* and can be manipulated to give bias value to a likelihood function. Thus, auxiliary assumptions must be justified to ensure a well-defined likelihood. I direct the reader to Appendix 1 at the end of the book, where I give further consideration to auxiliary assumptions and their justification and reflect upon those employed in my argument for human resurrection.

Bayesian Conditionalization

In further chapters, I will show that there is a considerable amount of evidence for and against the existence of God and it would be impossible to consider their confirmatory or disconfirmatory force on the proposition *p* all at the same time. So, each evidential element must be considered individually; however, after consideration of each one, the prior probability of the individual (subjective) or group (intersubjective) will change on the basis of it, (provided the initial prior value was not 0 or 1) and before consideration of the next evidential element this change must be accounted for. Revising priors in light of the evidence is only expected to occur if the evidence is actual rather than hypothesised. This change of prior probability caused by actual evidence is termed Bayesian conditionalization and it must be taken in to consideration in the cumulative procedure.

Cumulative Bayesianism

Consider the contorted theorem once again in light of the first evidential element e_1, the 1st prior probability $\Pr(p|k)^{1st}$ and the likelihood ratio λ_1:

$$\Pr(p|e_1 \,\&\, k) = \frac{\lambda_1 \times \Pr(p|k)^{1st}}{[\lambda_1 \times \Pr(p|k)^{1st}] + [1 - \Pr(p|k)^{1st}]}$$

and the likelihood ratio:

$$\text{Where } \lambda_1 = \frac{\Pr(e_1|p \,\&\, k)}{\Pr(e_1|\neg p \,\&\, k)}$$

If there is confirmation or disconfirmation, then a new prior probability will emerge, prior that is to consideration of the confirming or disconfirming force of the next evidential element e_2. Thus:

$$\Pr(p|e_1 \,\&\, k) = \frac{\lambda_1 \times \Pr(p|k)^{1st}}{[\lambda_1 \times \Pr(p|k)^{1st}] + [1 - \Pr(p|k)^{1st}]} = \text{BCONPr}(p|k)^{2nd}$$

Where BCON stands for Bayesian Conditionalization; the process then moves on to consideration of the next evidential element and so on:

$$\Pr(p|e_2 \,\&\, k) = \frac{\lambda_2 \times \Pr(p|k)^{2nd}}{[\lambda_2 \times \Pr(p|k)^{2nd}] + [1 - \Pr(p|k)^{2nd}]} = \text{BCONPr}(p|k)^{3rd}$$

$$\Pr(p|e_n \,\&\, k) = \frac{\lambda_n \times \Pr(p|k)^{n}}{[\lambda_n \times \Pr(p|k)^{n}] + [1 - \Pr(p|k)^{n}]} \text{BCONPr}(p|k)^{n+1}$$

Where e_n is the final evidential element

An example of how this can work in practice is shown below: see *Example* below.

Thus, the only data needed to work this process is the prior probability of *p*, which changes with Bayesian conditionalization as the analysis progresses and λ, which is determined from the ratio of the likelihood functions. Moreover, determination of λ is also a demand for reason rather than bias. *p* and ¬*p* are assumed in the nominator and denominator in the ratio λ respectively. Together with background knowledge *k*, the functions in λ request a view on how likely the evidence *e* is to be found. Hence, the atheist must, for the sake of the process, assume the truth of the proposition *God exists* in the case of Pr(*e*|*p* & *k*) and the theist must assume the truth of the proposition *God does not exist* in the case of Pr(*e*|¬*p* & *k*).

Independence

Bayesian conditionalization highlights the importance of the independence of the individual evidential elements, independence that is, in terms of their probabilistic interrelationship. For example, if the first evidential element e_1 entails e_2 then once Bayesian conditionalization has occurred on the basis of e_1 that evidential element then becomes part of the background knowledge *k* when considering the next evidential element e_2

$$\text{Where } \lambda_2 = \frac{\Pr(e_2|p \,\&\, k[\text{incl.}e_1])}{\Pr(e_2|\neg p \,\&\, k[\text{incl.}e_1])}$$

If e_1 entails e_2 then $\lambda_2 = 1/1 = 1$ and the differential force of *p* and ¬*p* is irrelevant, which is self-defeating.

Old Evidence

It could be argued that a Bayesian analysis based upon evidence already existing is intrinsically flawed on the grounds that $\Pr(e|k) = 1$ and $\Pr(e|p \& k) = 1$ and therefore $\Pr(p|e \& k) = \Pr(p|k)$; that is, the evidence can never confirm or disconfirm p and no Bayesian conditionalization can occur.[31] However, consider $\Pr(e|k)$ and $\Pr(e|p \& k)$, the probability of e is based upon the assumption of k and $(p \& k)$ respectively where k in both cases should not include knowledge of e as it is e that is under consideration.[32] The problem arises only when k is assumed to include e; thus, the subjective Bayesian should proceed in a way where the actual existence of e is assumed to be unknown. Notwithstanding, bias can creep in; for example, someone who has had a religious experience (e_R) may be inclined to give $\Pr(e_R|p \& k)$ a higher value and $\Pr(e_R|\neg p \& k)$ a lower value when determining λ than someone who has not had such an experience.[33] This is where intersubjective Bayesianism has its advantage over subjective Bayesianism; bias can wash-out through consensus. Nevertheless, the wise Bayesian must be aware of this lurking danger.

Example

A simple, hypothetical example may help clarify the form of Bayesian analysis I intend to adopt:

Returning to London from visiting to an aging aunt living in the country, I am making my way back, full of tea and Victoria sponge, to the local railway station to catch my train back to London Paddington. I'm walking along 400 yards from the station and I can make out the platform, but not the line; it crosses my mind whether I should break into a trot just in case a train is about to pull in to the station. The following evidential questions come to mind:

- Are other people running towards the station?
- Did I hear the words "Waterloo" in an otherwise incomprehensible announcement?
- Are there passengers waiting on the platform?
- Can I hear the distant rattling of train wheels?

The following background knowledge pervades my thoughts:

- I've missed my train so many times for dawdling.
- Trains not destined for Paddington stop at this station.
- There is a train to Paddington every 45 minutes.

Let:

- q be the proposition ‘the Paddington train is about to pull into the station.’
- e_1 be ‘people are running towards the station.’
- e_2 be ‘the word “…Waterloo …” was spoken on the station’s announcement system.
- e_3 be ‘there are people standing on the platform.’
- e_4 be ‘there is a discernible rattling of train wheels.’ (For the purposes of this demonstrative example, e_4 is to be taken as a mistaken perception and therefore not factual, whereas e_1, e_2 and e_3 are to be considered factual)
- k be the background knowledge stated above.
- My prior probability $\Pr(q|k)^{1st} = 0.4$ (a subjective value; more improbable than probable).

From the likelihood ratio $\Pr(e_n|q \ \& \ k)/\Pr(e_n|\neg q \ \& \ k) = \lambda$ for each evidential element, I make the following subjective decisions: λ for $e_1 = 2$; λ for $e_2 = 0.2$; λ for $e_3 = 2$; λ for $e_4 = 5$.

Let Bayes’ theorem be presented thus:

$$\Pr(q|e \ \& \ k) = \frac{\lambda \times \Pr(q|k)}{[\lambda \times \Pr(q|k)] + [1 - \Pr(q|k)]}$$

For e_1:

$$\Pr(q|e_1 + k) = \frac{2 \times 0.4}{[2 \times 0.4] + [1 - 0.4]} = 0.57$$

e_1 is *fact* therefore, $\text{BCONPr}(q|k)^{2nd} = 0.57$

For e_2:

$$\Pr(q|e_2 + k) = \frac{0.2 \times 0.57}{[0.2 \times 0.57] + [1 - 0.57]} = 0.21$$

e_2 is *fact* therefore, $\text{BCONPr}(q|k)^{3rd} = 0.21$

For e_3:

$$\Pr(q|e_3 + k) = \frac{2 \times 0.21}{[2 \times 0.21] + [1 - 0.21]} = 0.35$$

e_3 is *fact* therefore, $\text{BCONPr}(q|k)^{4th} = 0.35$

For e_4:

$$\Pr(q|e_4 + k) = \frac{5 \times 0.35}{[5 \times 0.35] + [1 - 0.35]} = 0.73$$

e_4 is *NOT fact* therefore, BCONPr$(q|k)^{5th}$ remains at 0.35

The cumulative subjective Bayesian process yields a probability of 0.35 of the Paddington train about to pull into the station: possibly a mild trot rather than a mad dash is called for. Also note as e_4 is not the case, no Bayesian conditionalization occurred and the final posterior probability $\Pr(q|e\ \&\ k)$ remains the same as for e_3. Normally, the e_4 calculation would be ignored.

VICARIOUS PROBABILITY

I wish to introduce a new idea into the domain of subjective probability; that is, vicarious probability, an extension of intersubjective probability introduced above.

Vicarious probability is a *subjective* process in the sense of involving the personal judgement of an evaluator or evaluators and *vicarious* in the sense that the evaluator(s) judge what probabilistic decisions an intersubjective syndicate would probably make. Thus, the evaluator(s) are predicting the syndicate's decisions based upon the mix of beliefs of the members comprising the syndicate. For example, a sports journalist might attempt to predict the decisions of a football selection panel based upon a knowledge of the beliefs of the members of the panel.

The Hypothetical Syndicate

Although the evaluator(s) are actual, the syndicate need not be; it could be hypothetical, reflecting a range of beliefs of the hypothetical members that comprise it. For example, the sports journalist may attempt to predict team selection based upon the expected decisions of a hypothetical selection panel comprising members holding differing beliefs concerning defensive and attacking strategies. The journalist would publish what he believes to be the best team based upon a conceived selection consensus that accounts for different justifiable strategies.

This is the approach I intend to adopt; that is; two actual evaluators considering the possible deliberations and decisions of a hypothetical syndicate comprising an atheist, an agnostic, and a theist. I have selected two evaluators;

an atheist and theist to mitigate evaluator bias. The evaluators will judge if this hypothetical syndicate would reach an intersubjective consensus or not on a set of probability values and likelihood ratios concerning the existence of God and in later chapters, human resurrection. Although this is a subjective process, it will be contained by Coherence through the application of probability theory.

Justification

Why a hypothetical syndicate rather than an actual one? Vicarious probability in this context needs justifying.

Even if the intersubjective process of an actual syndicate strictly complied with all the conditions of intersubjective probability and remained Coherent, there is a distinct possibility that different syndicates will reach different consensus values. However, I do not believe they would be wildly different; that is, most results would fall within a limited range. This would not be surprising, for if the process is probabilistically Coherent, complies with the intersubjective conditions and due consideration given to the evidence, different syndicates would truth-track to similar results. It is my view that if the vicarious approach also complies in a similar way and accounts for all the available evidence, I would expect it to yield a result similar to most actual intersubjective syndicates.

In the context of the argument for human resurrection there are both advantages and disadvantages to this vicarious approach as well as the adoption of a hypothetical syndicate as follows:

Double Counting

With the vicarious approach, prior knowledge by the syndicate members of the evidential elements can be assumed to be non-existent, whereas members of an actual syndicate would probably possess some prior knowledge of the evidence. This is liable to contribute to their prior probability values creating a risk of double counting in their analysis. For given such prior knowledge; that is, evidence already accounted for in the prior probability value, it may be accounted for again when that evidence is reassessed by the syndicate.

Evaluator's Knowledge

Unlike an actual syndicate, prior knowledge by the evaluators can be extensive. This has the advantage of increasing objectivity by granularity as opposed to the more general detail expected to be grasped by an actual syndicate meeting the evidence for the first time. Thus, a wide perspective of information, argument, and evidence can be assumed to be grasped by a hypothetical syndicate adding objectivity to the process.

Information and argument could be provided to actual syndicate members in advance, but this presupposes that each member would be able to allocate enough time to it and even then, there is no certainty that they would fully understand the material.

Stress

Emotional stress upon Bayesian conditionalization is avoided. Belief change can be stressful, particularly with an emotive subject like religious belief. A believer can easily become distressed when coming to terms with evidence that challenges their faith. This can lead to conflict within an actual syndicate, fracturing the all-important cohesion of the syndicate, and reducing the chance of consensus for the wrong reasons. Such potential destabilization is eliminated with a vicarious approach.

Time Limitations

Given the extensive evidential set and its complexity, time constraint would be a problem with an actual syndicate. A syndicate convening to consider, in depth, the entire evidential set related to the existence of God may need to meet for a long time. Syndicate members would undoubtedly have other commitments and convening such a lengthy meeting may prove difficult. Moreover, fully engaging with the consensus seeking process over a long period would require considerable concentration, and tiredness is liable to creep in. This can be mitigated with breaks and spreading the meeting over several days, but this would exacerbate the time limitation problem. These difficulties are avoided with a hypothetical syndicate.

Bias

Some members of an actual group may want to support or damage the research for reasons other than the force of the evidence under consideration; for example, to please the moderator or researcher or to proselytize. This may not occur, but the hypothetical syndicate is exempt from this danger. Clearly, there is also potential for bias with the evaluators, which I address below.

The disadvantages of vicarious probability are as follows:

Evaluator Bias

There is a risk of confirmation bias, whereby prior beliefs prejudice judgement of evidence and argument in favor of those beliefs.[34] However, reasons must be provided for the evaluators' judgments giving the reader the opportunity to adjust, but even then, the reader must be aware of their own potential bias.

One way I have mitigated such evaluator bias is to employ both an atheist and theist as evaluators. The prior probability of the evaluators will not count in their judgements, eliminating many syndicate-type dangers mentioned above and bias will be reduced if the evaluators can reach consensus on a vicarious judgement.

Loss of Syndicate Interaction

A vicarious approach loses the discursive interaction of an actual syndicate, thus reducing objectivity. By way of mitigation, this loss is made up by the greater in-depth knowledge of the evaluators, including argument and counter argument. Notwithstanding, there is some loss in the advantages of live syndicate interaction.

Background Knowledge

A fair-sized syndicate would have a considerable pool of background knowledge at its disposal. Evaluators would also have background knowledge which can be applied to the vicarious approach, but again there will be some loss.

Given the size, complexity, and nature of the probabilistic analysis of the propositions *God exists* and *dead persons will be resurrected*, the vicarious approach has considerable advantages over an actual syndicate approach.

Default to Subjective Probability

In those cases, where the evaluators believe that consensus is unlikely to reached, they should default to vicarious *subjective* probability; that is, individual values rather than syndicate consensus values. Justification of this *default to subjective probability* approach can be demonstrated through the Ramsey De Finetti theorem.[35]

Vicarious Priors

The first requirement of the evaluators in a Bayesian vicarious probability analysis involving a hypothetical syndicate is to select the prior probability values of the hypothetical members comprising the syndicate. These are the degree of beliefs prior to consideration of the relevant evidence, and because the members are hypothetical, the priors will be assumptions and those assumptions must be warranted.

Prior probabilities are degrees of belief in the truth of a proposition prior to knowledge of the relevant confirming and disconfirming evidence. The priors, in the context of this study, are propositional; that is, a statement believed by someone to possess a degree of truth-value on the probability continuum from >0 to <1, where 0 is believed to be certainly false and 1 is believed to be certainly true and, for example, a mid-point between 0 and 1 is believed to be just as false as true.[36] Note that, in the case of the application of Bayes' theorem, the probability range on the continuum must be from >0 to <1 as opposed to 0 to 1.

As I explained in the Introduction, atheism should be accounted for in any probabilistic investigation concerning the existence of God. Consequently, the hypothetical syndicate will comprise three hypothetical members; an atheist, an agnostic, and a theist and, to mitigate evaluator bias, one evaluator will be an atheist and the other a theist.

With an actual syndicate, the prior probability values held by its members are simple to determine by just asking them. However, with a vicarious approach there are no actual members to ask; any priors are assumptions made by the evaluators. Ideally, such assumptions would be informed by a formal principle for ascribing prior probabilities; for example, the principle of indifference. However, apart from being subject to paradoxes, the principle of indifference would only yield the agnostic's position on the probability continuum; a value true by definition.[37] Moreover, equal distribution of the probability continuum into quartiles for the atheist, agnostic, and theist would be an unjustified arbitrary principle.

The probability calculus assumes that probabilities are real numbers and each probability on the continuum should, theoretically, be represented by an infinite decimal (e.g., 0.1 is given by 0.999…∞) because the probability space between any one point on the continuum and another is infinitely divisible. However, with subjective and intersubjective probability the calculus is an approximation as values are vaguer. Given this, an alternative to the application of a formal principle would be to examine how the priors would be subjectively determined with an actual syndicate and extrapolate this to the vicarious context.

Selection of an actual syndicate that represents atheism, agnosticism, and theism would involve asking potential members whether they believe in the existence of God. This will result a range of responses; for example, "no I don't"; "yes I do"; "to some extent"; "sometimes I do, sometimes I don't," and many more possibilities. Bayesianism requires such subjective descriptions to be quantified on the probability continuum >0 to <1 and, to reduce interviewer bias, the respondents should evaluate their own responses. There are several ways for them to do this; for example, directly with the probability continuum itself, percentages, and fractions. Fractions do not, without complexity, correlate well with the probability continuum and direct correlation with the continuum, with its use of decimals, is not common practice and would probably cause confusion in the correlating process. Percentage correlation would be the most user-friendly process from 0% correlating with such expressions as "I'm absolutely certain there's no God" through to 100%, correlating with such expressions as "I've no doubt whatsoever that God exists."[38]

Given the vagueness of probabilistic descriptors such as "I'm not so sure, but I don't reject it out of hand," quantifying them in percentage terms would certainly be by whole number percentages: 1% or 15% or 90% and so on. This converts to two decimal place values on the probability continuum; 0.01 or 0.15 or 0.90 and so on. A percentage continuum comprising whole numbers produces 101 numbers from 0 to 100 to choose from which is bewildering and unwieldy. Most respondents would probably limit themselves to 5% increments; that is 0%, 5%, 10%, 15%...100%, although potential syndicate members with extreme views may choose smaller increments at the extremities of the continuum; for example, 1% or 99%. This would be to reflect their certainty, but at the same time acknowledge the slim possibility of a false belief.

Assuming 50% represents perfect agnosticism, then 40% and 45% represents *near agnosticism* in the atheism zone of the continuum and 55% and 60% in the theism zone. The range 0% to 10% in the atheism zone and 90% to 100% in the theism zone are low and high values respectively. Given that the potential members are not yet cognizant of the confirming and disconfirming evidential set relevant to the theistic proposition *God exists,* these low and high values may be indicative of a bigoted state of mind concerning the theistic proposition. This is not necessarily the case, and the choice of such values would not question the integrity of the believers, but provided there is a good population to select from, it would be better to avoid selecting members holding these prior values for the sake of constructive harmony in the syndicate.

Avoiding selection of an atheist and a theist holding prior-values in the near agnosticism zones, there are five values in the atheism and theism zones that potential members could fill:

Atheism—15%, 20%, 25%, 30%, 35%

Theism—65%, 70%, 75%, 80%, 85%

With the intention of maximizing conviction distance between atheism and theism to establish a clear belief-distinction at the outset of the syndicate's deliberations, 15% and 85% would be the ideal percentages to aim for. For the agnostic member, 50% would be the ideal percentage to aim for. Provided the population of potential members was large enough, these prior probability values could be found in that population.

The vicarious approach can assume such a state of affairs and the evaluators agree that the above percentages are the ideal for a hypothetical syndicate comprising three members; an atheist, an agnostic, and a theist. 15%, 50% and 85% translates into 0.15, 0.5, and 0.85 on the probability continuum and it is these priors that will be adopted by the evaluators.

CONCLUSION

The chapter shows that probability is a broader concept than number collation and statistics; it is also about what we believe to be true and subjective probability is a sound method of engaging with this facet of the concept. Provided the analyst remains within the axioms of probability and accounts for all the evidence including background evidence, their conclusions are warranted albeit subjective. Moreover, such warrant can be strengthened by intersubjective probability; a method of diluting any personal bias. Notwithstanding, subjective probability is a form of epistemic probability; that is, the rational degree of belief of an individual or group. Note that a rational degree of belief is more than a psychological state, it requires justification.

Bayes' theorem is derived from the definition of conditional probability and is a respected and widely used tool in comparative probability, both in its objective and subjective forms. Herein, I concentrate on its subjective and intersubjective application, the latter being a process I employ in a practical way through a vicarious approach; evaluator judgement on hypothetical intersubjective consensus formation.

From this chapter, the reader should now be aware that evidence is the engine of Bayesian confirmation theory and in the following chapters I intend to consider the generally accepted set of evidence relevant to the propositions *God exists* and *dead persons will be resurrected.* Such evidence is a mix of confirming and disconfirming evidence and it is important that none is missed.

NOTES

1. For the non-specialist but interested reader, I recommend Hacking I. (2001) for an excellent in-depth introduction to probability theory.
2. There are two theories of objective probability: the frequency theory and the propensity theory.
3. Epistemic probability is derived from the philosophical term epistemology, i.e., the theory of knowledge.

4. Belief is variable and can be viewed as a continuum from certainty through agnosticism to total disbelief.

5. Gillies D. (2003); p. 1f.

6. David F. N. (1992).

7. Laplace P. [1814] (1998); p. 4.

8. A theory of matter and energy at the subatomic level that is distinct from the cause and effect observable on the large scale. At this level, energy is absorbed and released in distinct quantities called 'quanta,' matter displays the qualities of both waves and particles and is probabilistic rather than deterministic in behaviour.

9. Keynes J. M. (1996); p. 56.

10. The principle of indifference relates to the application of a numerical value of probability to alternative propositions. Where the different values cannot be determined, the principle states that they should be allocated equal probability. For a discussion of the paradoxes that ensue from the application of the principle see Gillies D. (2003); p. 37ff.

11. An axiom in this context is a self-evident truth; therefore, an irreducible starting point for further analysis.

12. Swinburne R. (2004).

13. Kolmogorov A. N. (1933)—*Foundations of the Theory of Probability*—New York: Chelsea Publishing Co.

14. Gillies D. (2003); p. 59ff.

15. Ramsey (1926) and de Finetti B. (1974)—*Theory of Probability* (2 vols): New York: Wiley.

16. Gillies D. (2003); p. 58f.

17. A Dutch-Book bet can be as follows with P being the punter and B the bookie:.

   1. P bets on 4 horses in a multiple horserace as follows: Horse1: £100; Horse2: £50; Horse3: £40; Horse4: £20. Total sum bet = £210.
   2. B set the odds as follows: Horse1: evens; Horse2: 3 to 1 against; Horse3: 4 to 1 against; Horse4: 9 to 1 against.

   The devious bookie has accepted P's bets without warning P that a Dutch-book will occur with the sums placed; consider the following:.

   3. The betting quotients and probabilities are as follows: Horse1: 1/(1 + 1) = 0.5; Horse2: 1/(3 + 1) = 0.25; Horse3: 1/(4 + 1) = 0.2; Horse4: 1/(9 + 1) = 0.1. Sum of the probability values = 1.05.
   4. Sum payable to P for any one horse winning: Horse1: £100 + £100 = £200; Horse2: £150 + £50 = £200; Horse3: £160 + £40 = £200; Horse4: £180 + £20 = £200.

   P has laid-out £210, and no matter which horse wins, B only pays P £200: a Dutch-Book! If the probabilities had added-up to 1 or less, then P could be in front. For example, if horse4 is withdrawn and B returns £20 to P and does not adjust the odds, then the probabilities would sum to 0.95 and P would have only bet £190, so if one of the other horses had won P would be in front.

18. The difference between a definition and an axiom is that a definition expresses the meaning of an expression in this case 'conditional probability,' where as an axiom is an assumption assumed true without proof.

19. Gillies D. (2003); p. 62ff; Gillies refers to the definition of conditional probability as axiom 3: the multiplication law: Gillies D. (2003); p. 59.

20. Gillies D. (1991A); Agnaldo Portugal (2003) provides an insightful application of intersubjective probability to the philosophy of religion.

21. Gillies D. (2003); p. 169ff.

22. *Ibid*; p. 172.

23. Inductive logic refers to the formal system of inference from the particular to the general, echoing deductive logic, i.e., the formal system of inference by necessity from premises to conclusion. In the domain of probability, inductive logic is associated with Keynes [Keynes J. M.[1921], (1996); p. 241ff].

24. It was the good Reverend Thomas Bayes of Tunbridge Wells (1702-1761) who stated the theorem, although it was his colleague Richard Price who first presented it after Bayes' death to the Royal Society in 1763.

25. Howson C. & Urbach P. (1991); p. 26f.

26. Sometime referred to as the odds form of the theorem.

27. Conan Doyle (1893)—*The Adventure of the Final Problem.*

28. Conan Doyle (1903)—*The Adventure of the Empty House.*

29. Sober E. (2002); pp. 24–30 for further consideration of likelihoodism.

30. Proof (ignoring *k* for purposes of clarity):.

1. *From Bayes' theorem:* $Pr(p|e) = Pr(e|p)Pr(p)/Pr(e)$.
2. *From the theorem of total probability:* $Pr(e) = Pr(e|p)Pr(p) + Pr(e|\neg p)Pr(\neg p)$.
3. *From 1 & 2*: $Pr(p|e) = Pr(e|p)Pr(p)/Pr(e|p)Pr(p) + Pr(e|\neg p)Pr(\neg p)$.
4. *Equivalence*:.
5. $Pr(e|p)Pr(p) + Pr(e|\neg p)Pr(\neg p) = Pr(e|\neg p) \times [Pr(e|p)Pr(p)/Pr(e|\neg p) + Pr(\neg p)]$.
6. *From 1, 2 & 4*:.
7. $Pr(p|e) = Pr(e|p)Pr(p)/Pr(e|\neg p) \times [Pr(e|p)Pr(p)/Pr(e|\neg p) + Pr(\neg p)]$.
8. Let $\lambda = Pr(e|p)/Pr(e|\neg p)$.
9. *From 5 & 6 and axiom 1*: $Pr(p|e) = \lambda Pr(p)/\lambda Pr(p) + Pr(1—Pr(p))$.

31. Glymour C. (1980); p. 86.

32. Howson C. & Urbach P. (1991); p. 270f.

33. Bartholomew D. J. (2001); p. 18f.

34. Fischer J. and Mitchell-Yellin B. (2016); Ch. 12.

35. Let there be a horse race about to start and a bookie (B) and an evaluator (E) assessing the possible decisions of a hypothetical two-punter betting syndicate. The bet is between B and E and E's betting decision is based upon the possible decisions of imaginary punters P1 and P2 in a syndicate P.

1. B is betting against E on horse H winning.
2. E is representing P1 and P2 as a two-punter syndicate P albeit hypothetical.
3. *Assumption*: P1 would probably choose betting quotient $q1$.
4. *Assumption*: P2 would probably choose betting quotient $q2$.
5. If E bets with B on the basis of $q1$ and $q2$ she is proposing two quotients for the syndicate P.
6. *Assumption*: $q1 > q2$.
7. B chooses the stake $S > 0$ on the bet with E on the quotient $q1$ and stake $-S$ on the bet with E on the quotient $q2$.

8. H wins and B pays E the product of the stake and each quotient and returns the stake.
9. *From 8*: B's gain/loss = $q1$S − S + $q2$–S + S = S($q1$—$q2$) (the return of S is +ve for B as S was –ve).
10. H loses and B keeps the product of the stake and quotient.
11. *From 10*: B's gain/loss = $q1$S + q2–S = S($q1$ − $q2$).
12. *Conclusion*: with $q1 > q2$ B can Dutch-book E. Therefore, unless $q1 = q2$ in the syndicate P, it is better for E not to bet representing a syndicate, but representing two hypothetical individuals P1 and P2 and their individual but differing quotients $q1$ and $q2$.

The above demonstrates that an evaluator assessing the possible decisions of a hypothetical betting syndicate is better to default to subjective probability in those circumstances where consensus is unlikely. Where consensus is likely, then intersubjective probability is applicable. (Compare Gillies D. (2003); p. 170f).

36. Theoretically, such a mid-point would be 0.4999…∞.

37. "The principle of indifference asserts that if there is no *known* reason for predicating of our subject one rather than another of several alternatives, then relatively to such knowledge the assertions of each of these alternatives have an *equal* probability." Keynes J. M. (1996); p. 45.

38. Maintaining consistency in correlating such expressions with numeric values is fraught with difficulties: Mosteller F. & Youtz C. (1990).

Part II

UNDER STARTER'S ORDERS

Evidence for God

If I were to ask a zealous evangelical Christian if God exists, the reply would be "of course!" By contrast, if I were to ask an ardent atheist the same question, the reply would probably be "of course not!" The question of whether God exists or not is a central argument in the philosophy of religion and the foundation stone of my argument for human resurrection. This is because I am basing the formulation of the prior probability of my resurrection proposition (*dead persons will be resurrected*) on a probability network between that proposition and the theistic proposition *God exists* and the proposition *Jesus died and was resurrected.*

I do not believe anyone can prove God exists in the same way that I can prove Nelson's Column exists by, for example, catching a bus to Trafalgar Square looking at it and pointing it out to doubters.[1] However, there are things that we strongly believe exist that we cannot, as yet, prove; for example, dark matter. At the moment, there exists considerable circumstantial evidence that dark matter exists, but as the name suggests we cannot look at it; we cannot be absolutely certain. The existence of God is in that type of category in the sense that we cannot see God directly and some have in the history of mankind, then it is not an event that is easily repeatable. Therefore, it is a question of probability based upon evidence; the stronger the supporting evidence the greater the probability of the truth of the theistic proposition *God exists*. Likewise, the stronger the refuting evidence the less the probability of the truth of the theistic proposition. Both supporting (or confirming) evidence and refuting (or disconfirming) evidence must be accounted for. In fact, all relevant evidence must be accounted for to come to a sound probabilistic conclusion. Consequently, the following chapters and sections address the evidence that either confirms or disconfirms the theistic proposition *God exists*.

It is a vicarious process in so far as the results of the analysis are those envisaged by the evaluators of what the most probable conclusions would be of a hypothetical intersubjective syndicate.[2] Much ink has been spilt on the evidential elements relevant to the proposition *God exists* and I have selected what I consider to be the most relevant.

THE READER'S ASSESSMENT

There is an opportunity for the reader to undertake their own subjective analysis by evaluating the confirming or disconfirming force of each evidential element on the theistic proposition and to undertake a personal cumulative Bayesian analysis. In order to do so, the reader must first establish their prior probability of the theistic proposition before reading this chapter. Thus, before continuing, the reader must decide and record their degree of belief in the truth of the proposition *God exists*; from between 0 (certainly false) and 1 (certainly true) with 0.5 being just as probable as not. (Note that Bayes' theorem requires the values 0 and 1 to be excluded; see Chapter 2). Failure to do this prior to reading this chapter may distort their conclusions.

At the end of this chapter it is possible that the reader's degree of belief in the theistic proposition will change on the evidence; that is, Bayesian conditionalization. This probability value will represent their next prior probability of the theistic proposition as they consider the evidence in the following chapter. At the end of the three chapters the reader will have a posterior probability value of the theistic proposition on all the evidence provided herein. However, there is no reason why the reader cannot then add other evidence that they believe is relevant and undertake a subjective Bayesian analysis that will possibly lead to a further change in their degree of belief (see Appendix 2).

Methodology

The process of considering and analyzing the evidence through the lens of Bayesian confirmation theory introduced in Chapter 2 will apply to this and the following two chapters. The presentation of each evidential argument will follow a similar format as follows:

1. *Title and general description*: the conventional title given to the evidential element and its *prima facie* theistic/atheistic character.
2. *Confirming the evidence on the proposition*: this section will focus on the confirming force of the proposition on the evidential element under consideration. This will take the form of a likelihood function; that is,

assuming that God exists, how likely is the evidence in question to be discovered. This is expressed by the likelihood function $\Pr(e|p \,\&\, k)$, where e is the evidential element, p the theistic proposition *God exists*, and k the background knowledge. Background knowledge k is of vital importance in the analysis as will become clear further on.

3. *Confirming the evidence on the negation of the proposition*: This will be presented in terms of the likelihood function $\Pr(e|\neg p \,\&\, k)$, where $\neg p$ is the proposition *God does not exist*. Again, background knowledge k is of vital importance.
4. *Counter arguments and replies*: these will be arguments that challenge the main argument presented with responses by the author if appropriate.
5. *Evaluators' conclusion*: the evaluators will present a concluding statement based upon the arguments in support of $\Pr(e|p \,\&\, k)$ and $\Pr(e|\neg p \,\&\, k)$ respectively.
6. *Evaluator's judgement*: there are two evaluators; a theist and atheist and their function is to determine, by discussion, the likelihood ratio decisions of the hypothetical syndicate.

The likelihood ratio, first presented in Chapter 2, is as follows:

$$\lambda = \frac{\Pr(e|p \,\&\, k)}{\Pr(e|\neg p \,\&\, k)}$$

The value determined from the ratio will be rounded up to two decimal places to provide the λ-value, although a different λ-value can be determined by the reader if they so wish.[3]

The evaluators will attempt to argue for a consensus within the hypothetical syndicate. Where this is not possible, either fully or partially, then individual results will be allocated by the evaluators. This will represent a vicarious *subjective* probability value rather than an *intersubjective* one.

7. *Prior probability*: The prior probability of the proposition '*God exists*' is given by the function $\Pr(p|k)$ in the theorem and has been addressed in Chapter 2 and the following values selected by the evaluators: atheist: 0.15; agnostic: 0.50; theist: 0.85. These priors are liable to change after the Bayesian analysis of each evidential element. These new values then become the prior probability-value of the next element to be considered.
8. *Bayesian conditionalization*: this will also be hypothetical; that is, assumed to occur with the members of the hypothetical syndicate on the basis of whether the evidence is actual or related information indicates

that it is actual. Bayesian conditionalization will then be assumed to occur or not occur resulting in either three posterior probability values or none respectively. If conditionalization is assumed to occur, then these posterior probability values will be the prior probability values for the next evidential element under consideration. If there is no conditionalization, then the last prior probability values will be used for consideration of the next evidential element. When the evidential elements have been exhausted, the final result will be the vicarious subjective/intersubjective probability value of the theistic proposition '*God exists*' on all the evidence taken into account. I would urge the reader to review the process in Chapter 2 if the above is unclear.

9. *Posterior probability*: to remind the reader, the posterior probability is the probability of the theistic proposition *God exists* given the evidence and background knowledge and this will be determined from Bayes' theorem (likelihood ratio form) as presented in Chapter 2:

$$\Pr(p|e \,\&\, k) = \frac{\lambda \times \Pr(p|k)}{[\lambda \times \Pr(p|k)] + [1 - \Pr(p|k)]}$$

(λ is the likelihood ratio referred to above)

Once the likelihood ratio value λ is determined by the evaluators for each evidential element, it is entered into the theorem together with the three different prior probabilities values. Three posterior probability values will result and these values may converge or diverge as the cumulative analysis proceeds. These posterior values become the next priors for consideration of the next evidential element until the end of the evidential set is reached.

NOTES

1. The concept of 'proof' is usually applied to necessary truths such as those in mathematics and logic, so I'm using it in its common and looser sense; i.e., absolutely certain.
2. *Vicarious Probability*: Ch. 2
3. I noted in the process of determining λ by consensus, that specifying values for the two likelihoods helped reach agreement on the ratio value. Although technically unnecessary, I have included the likelihood values that the evaluators agreed upon.

Chapter 3

They're Off

Evidence from Divine Creation

The evidential elements under consideration in this chapter are those captured by the title 'evidence from divine creation.'[1] Note that the use of *divine* in this title is not intended to presuppose the outcome of the analysis. These elements are presented in sections and are known as the cosmological argument, the teleological argument, and the argument from providence. The relevant arguments for their confirming and disconfirming force are detailed below in each section as is the Bayesian process employed to determine that force.

§ SECTION ONE: THE COSMOLOGICAL ARGUMENT

The argument associated with this first evidential element has a long history connected to illustrious names such as Plato, Aristotle, Aquinas, Leibniz, Hume, and Kant; in fact, it was Kant that first introduced the term *cosmological argument*. It is an inferential argument; God's existence is deductively or inductively inferred by facts related to the existence of the universe. In the case of induction, such inferences are justified on causal grounds.

A correct deductive argument from evidence to hypothesis would ensure absolute certainty of God's existence and make consideration of further probabilistic confirming evidence otiose.

Confirming the Evidence on the Proposition

This relates to the likelihood function $\Pr(e|p \,\&\, k)$ where e is the proposition *the universe exists* (ignoring the fact that the universe already exists),[2] p the theistic proposition *God exists*, and k background knowledge. This likelihood function ascribes a probability value to e assuming p and given k. However, the usual confirming arguments on this topic relate to the posterior probability function Pr(*p*|*e* & *k*) directly; that is, given *e* and *k*, how probable is it that *p*. In both the likelihood and posterior functions *k* expresses the confirming or disconfirming force of the cosmological arguments.

A Causal Argument

The universe, being a contingent thing, requires an explanation of why it came about in terms of a cause distinct from itself; something cannot cause its own existence, for it would have to exist to cause itself. Contingent causes must themselves have a cause, therefore a contingent nexus of cause and effect must extend infinitely back in time, which is an infinite regress, defying any complete explanation or scientific explanation. However, cosmologists would argue that time began at the inception of the universe and cause presupposes time. If a necessary rather than a contingent thing was the first cause, then the regress stops and a complete explanation is possible. This could not be an explanation based upon the laws of nature given that such laws are based on the physics of our universe; therefore, argue some philosophers,[3] it must be a personal explanation pointing to God being the necessary first cause. Thus, the evidence confirming the existence of God is the universe itself.[4]

A Deductive Argument

It has been argued that the causal argument above is a deductive argument given that any contingent event must have a causal explanation; that is, caused by something other than itself; either another contingent event or a non-contingent event (a necessary cause). Given the *principle of sufficient reason*,[5] there must be a sufficient causal explanation of a contingent event. Because of the infinite regress, there is not a sufficient causal account from other contingent events. Therefore, from the *principle of excluded middle*,[6] there must be a non-contingent explanation; that is, a necessary one. What follows is a personal explanation in terms of an intentional act by God, as no natural event can be necessary.[7]

Counter Argument

An appeal to divine intentions seems problematic for a cosmological argument; that is, by observing the universe we conclude God exists or probably exists. However, if we point to the wonders of the universe and claim that this instantiation was what God intended, then we have assumed God's existence. To then conclude from this that because of the existence of the universe God exists or probably exists is to beg the question. The circularity can be diluted by claiming that it was the intention of a necessary being, but this is just a smoke screen. The circularity remains, as a necessary being with the wherewithal to instantiate the universe is what most think of as God. The cosmological argument could be halted at the divine intention stage, but the argument is not about intentions, it is about God's existence, requiring either a proof of existence or a probabilistic confirmation of existence.

Simplicity

An alternative argument by Richard Swinburne (2004) is an inference to best explanation stemming from the comparison of God with the universe in terms of simplicity. The argument proceeds by asserting that God is simpler than the universe, and that both God and the universe are scientifically inexplicable. In the case of God, a necessary eternal being, there are no antecedent causes and therefore, God is closed to any scientific explanation. Equally, any cause-effect nexus resulting in the universe cannot be explained from natural laws as these presuppose laws within the universe; thus, the universe, other than a divine explanation, is a brute fact.

Based upon the premise that simple, inexplicable things are more liable to exist than more complex inexplicable things, it is argued that the probability of the existence of God is greater than the probability of the existence of the universe, assuming a counterfactual state of affairs where neither God nor the universe exist for sure.[8] The argument proceeds by suggesting God would have good reason to instantiate the universe and that God's creative action is accounted for by a personal explanation in terms of conscious intentions; that is, a complete explanation requiring no further analysis.[9] Such explanation by intention is to be contrasted with a brute fact assertion. Given the principle of simplicity, the better explanation, argues Swinburne, is the God-cause rather than brute fact. Thus, the existence of the universe probabilistically confirms the existence of God.[10]

Counter Argument

Swinburne's *a priori* principle of simplicity is questionable, particularly when a logical theory of probability is based upon it; there is a regression into subjectivism when applying the principle. This is because there are no strict criteria for application; what counts as simple in one context does not in another.[11] There are no necessary and sufficient conditions for the application of the criterion; the application is variable between contexts and is ambiguous:

> Do we prefer simple theories merely for aesthetic reasons, or on pragmatic grounds, because such theories are easier to grasp, to handle in calculations, and to remember? Or should we prefer simpler theories because they are better testable

> or falsifiable, as Karl Popper argued? Or, finally, do we select simpler theories because their simplicity is somehow related to truth? [Philipse, 2014, 214f]

Daniel Howard-Snyder (1996) challenges the principle of simplicity by arguing that a world comprising nothing is simpler than a world comprising something. On the criterion of simplicity, nothing-world is more liable to exist than the infinite series of something worlds.

Reply

A particular world in an infinite series is incredibly unlikely to be the one that exists; there appears to be a contradiction.[12] This argument only works if all the worlds have an equal chance of existing. To claim, as Howard-Snyder does, that all worlds in the series have an equal chance of existing is to assume the principle of simplicity is false prior to the analysis.

More Simplicity

Swinburne does provide a set of 'facets of simplicity,'[13] but their application to the differentiation between God and the universe is questionable. In describing the complexity of the universe, Swinburne states:

> A complex universe…is indeed a rather complex thing…There are lots and lots of separate chunks of it. The chunks each have a different finite and not very natural volume, shape, mass, etc.—consider the vast diversity of the galaxies, stars, and planets, and pebbles on the seashore. [Swinburne, 2004, 150]

This detailed description contrasts with that of God:

> …the postulation of a God of infinite power, knowledge, and freedom is the postulation of the simplest kind of person there could be. God is an unextended object, the divine properties fit together, and they are properties of infinite degree;…infinite degrees of a property have a simplicity lacked by large finite degrees of the same property. [*Ibid*, 150f]

Reply

It is much easier to provide granular explanations about something there is access to and physics, astronomy, and cosmology have good access to the constituents of the universe and the governing laws therein. As the opening paragraph of Chapter 1 makes clear, God is not as accessible as the material world of the universe. Consequently, divine descriptors are limited to definitional predicates like *omnipotence* or *unextended*. It is no wonder the postulation of a God is simple, but it is not God that is simple, but the postulation, given there is very little we can say. Indeed, the less granular any description becomes the simpler it will appear.

Similarly, Swinburne makes a comparison between the maximal qualities of the omni-predicates with that of infinity and zero and argues that just as infinity and zero are simple so are the omni-predicates.[14] However, the set theory mathematician Georg Cantor (1845–1918) demonstrated that infinity is not as simple as first impressions may suggest, but is expressible in an infinite number of possible sizes for an infinite number of sets. Thus appearances of simplicity, because of descriptive restriction, do not entail actual simplicity.

Turning to the facets of simplicity, the first five have application in the theistic context (the sixth relates to mathematical formulation). These relate to a hypothesis and its simplicity based upon the following elements: the number of things postulated, the number of kinds of things postulated, the possibility of direct observation, governing laws and variables within the laws respectively. This is not to claim that the structure of the hypothesis is what is being assessed but its referent.

The omni-predicates are anthropomorphic formulations based upon what is meant by God; they are not observed characteristics like those of the universe. The fact is, if the criterion of simplicity has any credence in the cosmological argument, then access to data has to be equitable and it is not. Thus, none of the five facets can be justifiably applied to the God hypothesis on the grounds of restricted access; indeed, it is possible that God is much more complicated than the universe or, given a pantheistic vision, is the universe.

Swinburne's application of the principle of simplicity is one of *ceteris paribus*, that is to say "…other things being equal, a simpler hypothesis is more probably true [when] at work in choosing among scientific theories of equal scope fitting equally well with background evidence and yielding the same data with the same degree of logical probability." [Swinburne, 2001, 82f] However, given this variety of parameters, competing theories are unlikely to equalize enough to apply the deciding factor of simplicity.

Swinburne also appeals to the history of science in support of the criterion of simplicity, claiming that, in the past, simple scientific theories have been more successful.[15] However, there is a serious problem with such an appeal:

Let:

1. E be the evidential claim that simple historical theories have been more successful.
2. H be the hypothesis that simple theories are more successful.
3. H′ be the hypothesis that E is true, but will not continue to be true in the future.

4. *From 1, 2 & 3*: Pr(E|H) = Pr(E|H′) = 1
5. *From 4 & Bayes' theory in its comparative form (without k for clarity)*:

$$\frac{\Pr(H|E)}{\Pr(H'|E)} = \frac{\Pr(H)}{\Pr(H')}$$

6. *From 5*: Pr(H|E) > Pr(H′|E) if and only if Pr(H) > Pr(H′)

Thus, on the evidence E, if simple theories are more liable to be successful than less simple ones, it is because of the prior assumption (independently of E) that simple theories are more liable to be true. Consequently, Swinburne is begging the question.[16]

Simplicity may have application in subjective probability as a refining top-up to probabilistic decisions,[17] but as an *a priori* principle in logical probability it is left wanting.

A Modal Argument

There is a cosmological argument in terms of possible world modal logic. A contingent proposition is true in some possible worlds and false in others, whereas a necessary proposition is true in all possible worlds. In Chapter 1, I introduced a modal logic axiom termed S_5, which states that if a proposition is possibly necessary, it is necessary. Thus, given this axiom, consider the following:

1. If it is possible that a necessary supernatural universe-creator exists, then it is necessary that that a supernatural universe-creator being exists.
2. It is possible that a necessary supernatural universe-creator exists.
3. It is necessary that that a supernatural universe-creator being exists.

Given S_5, the argument depends upon premise 2.

Possible worlds contain necessary and contingent propositions; the former is true in all possible worlds, the latter is true in some but not all worlds. Thus, accepting that all the necessary true propositions can, hypothetically, be expressed as one proposition comprising a conjunction of all necessary propositions (the necessary conjunct), they would be true in *all* possible worlds. Likewise, in *each* possible world, the total contingent propositions therein can, hypothetically, be expressed as one proposition comprising a conjunction of all contingent propositions (the contingent conjunct).

All worlds contain the necessary conjunct, but no world contains the same contingent conjunct, which individuates the different worlds. Thus, if two possible worlds manifest the same contingent conjunct, they would be the same world.

The Weak Principle of Sufficient Reason

I referred to Leibniz's principle of sufficient reason above, but there is a weaker version in terms of possible world semantics.

Let: p be a proposition
q be a proposition
w be any world
w_1 be a possible world

Then the weak principle of sufficient reason states that if p is in the contingent conjunct of w, there is a possible world w_1 such that it contains p and q and 'q explains p' in its contingent conjunct. Hence, *sufficient reason* can be defined in terms of possibility rather than necessity.

The argument remains in the domain of possibility as follows:

From the weak principle of sufficient reason, it is *possible* that the contingent conjunct in the actual world is identical to the contingent conjunct in w_1. In w_1, there is also a proposition q that explains its contingent conjunct. Thus, for the actual world to be identical to w_1 it must also contain q.

The contingent conjunct in the actual world would contain either q or $\neg q$ but not both; however, the contingent conjunct in w_1 does contain q, but it also contains the actual world's contingent conjunct. If the actual world's contingent conjunct contains $\neg q$ then there is a contradiction; thus, the actual world must contain q.

This raises the question of what can explain the contingent conjunct; that is, the nature of q.

q is a contingent proposition, so forms part of the contingent conjunct. Consequently, q must report something beyond itself that explains the contingent conjunct, but does not form part it. A scientific explanation will not do as it comprises contingent propositions that would be part of the contingent conjunct. q could provide a personal explanation of the contingent conjunct in terms of the intentional action of an all-powerful being. Such a being could either be contingent or necessary, but if contingent there would have to be a proposition in the conjunct reporting His existence. This is a problem for the explanatory power of q, as it could not explain this existential proposition; existence is simply a presupposition of a being that intentionally acts. To eliminate this existential proposition in the contingent conjunct, the being would have to be necessary; no explanation of existence in the actual world or any possible world is therefore required. Perhaps this argument provides justification to premise 2 above and vindication of the S_5 cosmological argument.

It is suggested that this argument *justifiably* weakens the strong principle of sufficient reason. The strong version is of the form 'for any true proposition there is a further proposition that explains it,' the weaker version is of the form 'for any true proposition *it is possible that* there is further proposition that explains it.'[18]

Counter Argument

In Chapter 1, I rejected the logical necessity of God's existence. If necessity is definable as true in all possible worlds, then necessity dilutes omnipotence. If God is omnipotent, then God has the wherewithal to abstract himself from any possible world. However, the cosmological argument assumes necessary existence; that is, non-contingent existence.

John Mackie (1982) provides an interesting counter argument to divine necessity. If God necessarily exists, then He exists in all possible worlds, thus He cannot exist in some but not others. Such necessary existence is logically compatible with not existing at all; that is, either God exists or He does not exist,[19] but if He does, He exists in all possible worlds. This makes God a logically contingent being, but across all worlds rather than some and if we accept that God does exist, it raises the question of why He exists rather than not. This undoubtedly is of the brute fact domain, but does raise some doubt over the superiority of the divine intention argument (a necessary being) over the scientific explanation (contingent cause) of the universe.[20]

Confirming the Evidence on the Negation of the Proposition

This relates to the likelihood function $Pr(e|\neg h \ \& \ k)$; that is, assuming God does not exist, how likely are we to find the evidence? Again, the Bayesian must rely on the background knowledge k in the form of argument for a viable alternative cause of the universe other than God.

It could be argued that the universe does not require an explanation in terms of a cause; it just is a brute fact. Causal explanation results from the observation of causal relationships between things, but this is not possible with the universe. The concept of cause does not apply.[21] This because time did not exist at the point of instantiation and the concept of causation presupposes time; that is, an event or thing *prior* to the instantiation. In that sense, God cannot be said to *cause* the universe to exist.

We can and do apply the notion of causation to states of affairs without observing a series of repetitions of the same state of affairs. Not observing a series of universes being caused does not mean we cannot apply 'causation' to its inception.[22] Also an infinite regress of explanation is not obligatory, provided a high-level, rather granular explanation is sufficient. Even within our universe, explanation becomes pedantic at some point where no further explanatory gain is made. Swinburne refers to such excessive causal explanation as the *completest fallacy*.[23]

Inflation and string theory do provide an explanation, albeit mathematical; however, there is currently no scientific explanation, but this does not mean there never will be. Pre-universe scientific explanation is not closed even though it is an event outside time. Scientific explanation may have to embrace a non-causal paradigm. Science can engage with past events on the basis of extant evidence, likewise it can engage with states of affairs that resulted in the instantiation of the universe from extant evidence. Cosmological mathematics can provide some theoretical explanation for the instantiation of the universe and point to possible extant discoverable evidence for the theories; for example, microwave background radiation, space-time inflation, dark energy, and quantum field fluctuation. Indeed, inflationary theory for the instantiation of our universe and other universes, although thin on supporting evidence, does have some, if not quite enough for it to enter into the scientific canon at this time. Because scientific evidence is wanting does not entail it always will be; indeed, to exclude a possible future scientific paradigm in favor of an extant one in philosophical explanation is too restrictive.[24] To claim that a scientific explanation requires reference to scientific laws in the current paradigm and no other is to close future possibility, a future that is already on the horizon.

Counter Argument

The objection to a cosmological explanation has been touched upon above. It remains a mathematical postulate with some evidential support, but not enough to provide acceptance into the scientific canon. Moreover, if the principle of sufficient reason is adopted, then cosmology is unlikely to provide a full, complete, and ultimate explanation of the existence of the universe because the causal nexus resulting in its creation is liable to be limitless or even non-existent.

Reply

I argue above, that the principle of sufficient reason is too strong and a higher-level explanation rather than granular explanation would suffice, thus avoiding an explanatory infinite regress. Further, evidence in support of theoretical cosmology is liable to materialize as technical developments in this domain progress.

At present, there is a gap in scientific explanation and some fill that gap with a divine explanation. Perhaps it is better to view the two positions as compatible rather than incompatible and distinct. It is concerning that theistic explanation seeks stable ground, where science falters, and as soon as science catches up, theism moves on to the next gap in scientific explanation. This may well be the case with the cosmological argument; because a scientific explanation for the instantiation of the universe is wanting, then a divine explanation is called upon. Denial of the possibility of a scientific explanation for the instantiation of our universe is, in my view, a denial too far. If there is a God, then that being is intimately woven into the fabric of reality and the science thereof; it is foolhardy to seek theistic refuge where science has yet to establish itself. Rather, theism should embrace scientific explanation, with each new discovery viewed as a further clarification of the miracle of divine creation.

> Truth is indivisible: no genuine truth can conflict with any other truth; so if there *are* truths or religion, for example about transcendent reality, those truths must necessarily be compatible with any of the truths discovered by science. The religious believer thus has no business taking issue with the genuine discoveries of science, but on the contrary should warmly welcome them as illuminating and enriching our understanding of reality as a whole. [Cottingham, 2015, 35f]

In contrast, the atheist is also liable to embrace scientific explanation, not as woven into divine creation, but rather as an exhaustive explanation.

EVALUATORS' CONCLUSION

The cosmological arguments for a divine causation of the universe are subject to considerable objections to the point that it is difficult to give the likelihood function $\Pr(e|p \mathbin{\&} k)$ a high value on the basis of them. The only mitigating factor is not to be found in the cosmological arguments; that is, *from evidence e to God's existence p* (posterior probability), but in the very nature of God; an omnipotent, omniscient, and omnibenevolent being. Even rejecting the principle of simplicity, it would not be surprising that, if such a being exists, He would employ His enormous powers and love to instantiate a universe that would permit persons to evolve and express their creative free-will and love as a reflection of the divine character; that is, *from God's existence p to evidence e* (likelihood). Indeed, the very omni-predicates that define God ensure creation; not as a logical necessity, but in terms of a high probability. This argument echoes the auxiliary assumptions concerning the implications of the omni-predicates residing in the background knowledge k. (See Appendix 1) Thus, given that the likelihood function $\Pr(e|p \,\&\, k)$ assumes the existence of such a being, a high value by the hypothetical syndicate would be warranted. Notwithstanding, the likelihood function $\Pr(e|\neg p \,\&\, k)$ also has credence on the grounds of a naturalistic cosmological argument; that is, the mathematics of inflation and string theory provide a non-divine explanation for the existence of the universe with some supportive empirical evidence, albeit awaiting scientific recognition. At this stage in its development some explanatory skepticism must obtain, but a fair value would probably be given to the likelihood function $\Pr(e|\neg p \,\&\, k)$ by the hypothetical syndicate.

Evaluators' Judgement

Let:

e be the proposition *the universe exists*
p be the proposition *God exists*
k be the background knowledge related to the cosmological argument

1. *Atheist's view*: reluctantly, the atheist will give a high value to $\Pr(e|p \,\&\, k)$ on the required assumption of p and the divine attributes as an auxiliary assumption in k. (See Appendix 1) However, the atheist would also give a high value to $\Pr(e|\neg p \,\&\, k)$ on the auxiliary assumption that that science will eventually provide an exhaustive explanation of why the universe came about. Notwithstanding, the atheist will acknowledge that $\Pr(e|p \,\&\, k) > \Pr(e|\neg p \,\&\, k)$ on the grounds of the current weakness of an exhaustive scientific explanation.
2. *Agnostic's view*: the same as the atheist, but less value given to $\Pr(e|\neg p \,\&\, k)$.
3. *Theists view*: the same as the agnostics but with much less value given to $\Pr(e|\neg p \,\&\, k)$.

Consensus

An expected consensus between the agnostic and theist as follows (see footnote 3):

$\Pr(e|p \,\&\, k) = 0.7$
$\Pr(e|\neg p \,\&\, k) = 0.4$
$\Pr(e|p \,\&\, k)/\Pr(e|\neg p \,\&\, k) = \lambda = 1.75$

However, the atheist would be unlikely to compromise, preferring a subjective λ-value as follows:

$\Pr(e|p \,\&\, k) = 0.7$
$\Pr(e|\neg p \,\&\, k) = 0.6$
$\Pr(e|p \,\&\, k)/\Pr(e|\neg p \,\&\, k) = \lambda = 1.17$

Prior Probability

The following priors apply to this evidential element: atheist: 0.15; agnostic: 0.5; theist 0.85.

Bayesian Conditionalization

The existence of the universe is clear; therefore, it is reasonable to assume conditionalization by the hypothetical syndicate members. The posterior probability values below then become the priors for the next evidential element.

Posterior Probability

From Bayes' theorem:

$$\Pr(p|e \,\&\, k) = \frac{\lambda \times \Pr(p|k)}{[\lambda \times \Pr(p|k)] + [1 - \Pr(p|k)]}$$

Substituting the three prior probabilities of the hypothetical syndicate members into the theorem:

Atheist

$$\Pr(p|e \,\&\, k) = \frac{1.17 \times 0.15}{[1.17 \times 0.15] + [1 - 0.15]} = 0.17$$

Agnostic

$$\Pr(p|e \,\&\, k) = \frac{1.75 \times 0.5}{[1.75 \times 0.5] + [1 - 0.5]} = 0.64$$

Theist

$$\Pr(p|e \,\&\, k) = \frac{1.75 \times 0.85}{[1.75 \times 0.85] + [1 - 0.85]} = 0.91$$

§ SECTION TWO: THE TELEOLOGICAL ARGUMENT

This second evidential element is primarily associated with Hume, albeit as a refutation, but has been and still is, seen as a powerful piece of artillery in the theistic arsenal. It is allied to some extent with the cosmological argument and as such creates a problem for the Bayesian as will become clear further on.

Confirming the Evidence on the Proposition

Like the cosmological argument, the teleological argument as a probabilistic argument is presented in a posterior probability form; that is, the probability of the proposition '*God exists*' on the evidence of certain phenomena of nature. Again, such argument is contained within the background knowledge k in the likelihood ratio method employed herein; that is, $\Pr(e|p \;\&\; k)/\Pr(e|\neg p \;\&\; k)$ where e is the teleological evidence and p the proposition '*God exists*.'

Analogy

Looking at the rotatory motion of a bacterial flagellum, which gives rise to motion in the bacteria, the movement of a propeller comes to mind. It is but a short step from that *analogy* to the idea of intelligent design in nature. Indeed, if the flagellum and propeller share certain properties and the propeller exhibits those properties because it is designed by an engineer, then similarly, the argument goes, the flagellum is also designed by an engineer on the grounds that same effects have same causes. The former engineer is human, the latter God. This is an example of the natural theological argument by analogy; a teleological argument for the existence of God.[25]

Counter Argument

Hume argues that the analogy is limited; for example, being part of a living organism, the flagellum is self-sustaining, but the propeller is neither living nor self-sustaining. Moreover, the move from 'engineer' to God is questionable, given that much of nature is the cause of natural evil. For example, river blindness caused by a parasitic worm gives rise to horrendous human suffering, which is hardly the handiwork of an omnibenevolent God.[26]

It is the property analogy that is the key to this early teleological approach; properties such as complexity, functionality, aesthetics, efficiency, simplicity, and so on. We highlight these properties in nature because they are so central to human design: the argument from analogy is highly anthropomorphic. Indeed, there are natural properties that are not analogous to human design, such as reproduction and mutation.

Self-Evidence

Primitive cave paintings are self-evidently mind generated, and perhaps there are non-analogical properties in nature where creative mind is equally self-evident; for example, the double helix of the DNA molecule. Anthropomorphism is diluted by such a non-comparative approach. The self-evidence of William Paley's watchmaker argument is of this ilk, where he argues that a watch, compared to a stone, is self-evidently designed. Equally, he argues that a biological eye is self-evidently designed through analogy with a telescope.[27]

Counter Argument

Darwin's evolutionary theory by natural selection provides a compelling argument against the self-evidence argument. It is the causal nexus of nature that, over vast tracts of time, results in the apparent design of biological entities. Indeed, persons' perception of the world is incongruous with the enormous periods of time needed for natural selection to be evident on a grand scale, resulting in the false belief of self-evident intelligent design.

Fine Tuning

Given the adoption of evolutionary theory into the scientific canon, most divine design apologists have moved their focus from a granular explanation of certain biological examples, to a high-level explanation of indirect intelligent design; that is, making things make themselves. Thus, such apologists view the universe as being instantiated by God in such a way that its essential features bring about cosmic order and life. In particular, focus is now on the laws of nature that are so finely tuned that the cosmos and life therein emerged as a consequence. Such fine-tuning is demonstrated in the values of the many constants such as the strong and weak nuclear forces, the electromagnetic force, and gravity. Should such forces have exhibited slightly different strengths, life would never have emerged.[28] A probabilistic approach is frequently adopted with the fine-tuning argument in support of intelligent design, whereby it is argued that the existence of the finely tuned laws of nature is more probably caused by God than by something else such as chance or material causation. Swinburne employs Bayesian confirmation theory based upon the criterion of simplicity to this end.[29] Swinburne's approach is abductive; that is, an inference to best explanation, and his argument hinges upon God's intentions in creation.[30]

Counter Argument

The most compelling counter argument to that of divine fine tuning is that of multiverse cosmology whereby the possibility of an infinite multiverse exhibiting all the possibilities of natural laws explains why the laws of our universe are as they are. Combined with the anthropic principle, multiverse cosmology provides a persuasive naturalistic explanation.

Assumptions about divine intentions are extremely subjective and anthropomorphic, making it very difficult to assigned a probability value to Pr(e|p & k): (See Appendix 1).

> There are as many likelihoods as there are suppositions concerning the goals and abilities of the putative designer. Which of these, or which class of these, should we take seriously? [Sober, 2003, 38]

Confirming the Evidence on the Negation of the Proposition

I present three arguments that provide alternative naturalistic explanations for the existence of apparently designed biological phenomena and for the finely tuned laws of nature.

Evolution

As argued above, Darwin's theory of natural selection and the subsequent scientific developments in the theory of evolution provide a persuasive and alternative argument to the analogy and self-evident arguments for intelligent design. Indeed, there is probabilistic confirmation of the theory by evidence, and the theory possesses explanatory power, scope, plausibility, and predictability. In short, it eclipses both the analogy and self-evident explanations for intelligent design. Admittedly, there are some unexplained links in the evolutionary chain, but the absence of evidence does not entail that evidence will never be forthcoming; science marches forward and these gaps are expected to be filled.

The retreat from failed granular explanation for intelligent design by analogy to a high-level explanation of the finely tuned laws weakens, in probabilistic terms, the intelligent design argument; it is a God of the gaps shift. Consider the following explanatory shift:

1. *Assumption*: Jones burgled the house
2. *Conclusion from 1*: Jones is a thief
3. *Premise*: Jones was in the pub at the time of the burglary
4. *Conclusion from 3*: Jones did not burgle the house
5. *Assumption*: Jones arranged for someone else to do the burglary
6. *Conclusion from 5*: Jones is a thief

The assertion that Jones is a thief is maintained by shifting the argument from a more definitive assumption to more circumstantial one. There is a parallel in the move from the analogy/self-evident argument to that of the making-things-make-themselves argument of fine tuning to preserve the intelligent design assertion. Such a shift is also indicative of prior commitment for reasons other than the evidence; for example, the guilt of Jones despite evidence to the contrary.

Counter Argument

Neo-Darwinian evolutionary theory; that is, a contemporary approach to Darwinian theory in terms of genetic mutations, is not without challenge. This is particularly the case with macro-evolution or the emergence of new species compared to micro-evolution or mutational changes within a species. Moreover, evolutionary theory presupposes the existence of life for evolution to take effect; however, the emergence of life on Earth may defy a persuasive scientific explanation.[31]

Even if the evolution argument provides a convincing alternative explanation for certain biological phenomena that appear to exhibit intelligent design, it does not explain the fine-tuning argument for intelligent design.

Cosmology and Multiverse Theory

The mathematically orientated discipline of cosmology has become a prominent scientific enterprise, revealing many hidden mysteries of the universe and more recently explanations of its causation. Of particular note is inflation theory associated with Alan Guth, Andrei Linde, Paul Steinhart, and Andreas Albrecht, whereby a region of space, without our universe in it, is infused with an inflation field that, through negative gravity,[32] causes a peak of inflationary potential energy. That peak collapses due to quantum jitters or fluctuations, causing a massive and rapid inflation of space so much greater than the speed of light [33] and the creation of particles that eventually cool and coalesce to form the universe. This inflation process could conceivably occur in other regions of space causing a vast or even infinite number of other universes: the multiverse.[34]

Some universes potentially exhibit different laws to our universe, although the variations would be finite in number. Based upon the anthropic principle (see below), it is unremarkable that our particular universe has the laws it does rather than others, for life could not exist if they were different. Indeed, another universe with the same laws as our universe could also exhibit life, and if the multiverse were infinite there would be an infinite number that exhibit life.

Comparing the probabilities of intelligent design or chance as the cause of the finely tuned laws of our universe, chance surpasses intelligent design given the multiverse scenario with its vast combinations and permutations of laws together with the anthropic principle. Indeed, if there are an infinite number or even a finite number of universes that exhaust the possible combinations of laws,[35] then at least one universe will exhibit the laws observed in our universe, and that universe is our universe.

Counter Argument

Despite the underlying mathematical rigor, multiverse cosmology has not, as yet, been accepted into the scientific canon, on the grounds that substantial evidence is not forthcoming.

Reply

There is some supporting evidence for inflation theory such as the cosmic microwave background and accelerating expansion of space, and there is also the potential for evidence of a multiverse from the CERN research with the Large Hadron Collider.[36] Moreover, mathematics may be discovered rather than invented; that is, it is part of the fabric of reality and in this sense it is discovered akin to scientific discovery. Thus, mathematical cosmology may be considered as *evidence*, if not strictly scientific evidence, for the existence of a multiverse.[37]

The Anthropic Principle

The weak anthropic principle[38] states that the finely tuned laws of nature of the universe are deemed as essential given that mankind is now aware of them, as mankind's very existence depends upon such laws. Thus mankind's existence entails these laws of nature and hence such laws are unremarkable; they have to be just as they are. Thus, it is argued that assigning a higher probability to intelligent design than to just chance on the basis of the evidence of the finely tuned laws of nature is wrong. Rather, $\Pr(L|D) = \Pr(L|C)$, where L is a conjunction of propositions expressing the existence of certain finely tuned laws, D the proposition that the universe is intelligently designed and C a proposition that the universe came about by chance. In both the design and chance causation scenarios, observers emerge and it is they that observe L and ascribe the probability values. Hence, there is equality of the probability of L on D and L on C, given that observers entail L.[39]

> If you are alive and check to see whether the physical constants fall in the narrow range that permits life to exist, you are bound to find that they do, regardless of whether the constants had their values assigned by an intelligent designer or by a chance process. [Sober, 2009, 79]

Counter Argument

The existence of observers did not cause the universe with its finely tuned laws to come about, and this assignment of equal probability values is an 'after-the-event' assignment; that is, after observers have emerged. There is a parallel with the problem of old knowledge introduced in Chapter 2, whereby old knowledge of the evidence *e* in the likelihood function Pr(*e*|*p* & *k*) resides in the background knowledge *k* ensuring a probability value of 1 to the likelihood. Accounting for old knowledge in *k* does not permit the analyst to capture the probabilistic force of *p* on *e*. To capture the probabilistic force of D and *C* on *L* respectively requires a stretch of imagination to either a counterfactual context of no observers or a point in time prior to the emergence of observers and in both cases, design is much more feasible than chance to produce specific laws of nature.[40] Notwithstanding, it could be argued that this counterfactual/bygone approach infringes the principle of total evidence (see Chapter 2); that is, it is important to account for all the evidence in conditional probability and the existence of observers is an important evidential element as background knowledge *k* in the likelihoods Pr(*L*|*D* & *k*) and Pr(*L*|*C* & *k*). However, in the bygone scenario there were no life forms or observers.

An alternative, but related approach is based upon the 'unremarkable' assertion given the existence of observers. Swinburne (2004) provides an excellent analogy involving a dastardly card dealing machine that randomly deals a hand to a kidnapped victim. Unless a very improbable hand is dealt, the machine will explode killing the victim. The machine issues the cards, the very improbable hand results and no explosion occurs. The victim is pleasantly surprised and seeks an explanation. Notwithstanding, the kidnapper exclaims "why such surprise, it is quite unremarkable, for if it wasn't that hand you wouldn't be here!" However, the event is remarkable as are the finely tuned laws of nature and both occurrences cry out for an explanation.[41]

Double Counting

One serious problem arises with the teleological argument when it is employed in a Bayesian cumulative argument that also includes, as one of its evidential elements, the existence of the universe; that is, the cosmological argument. The cosmological argument, in its likelihood form, argues that the proposition *God exists* probabilistically confirms the evidence of the existence of the universe. The universe comprises, *inter alia*, the laws of nature that are so finely tuned as to enable the universe as a whole to exist and develop. Thus, for the universe to exist at all then the laws must also exist, yet the laws exist as a consequence of the universe existing; they are necessary not contingent. In short, the laws are an essential property of the universe and any definition, in terms of the essential properties of the universe, must include the laws. Therefore, to probabilistically confirm the existence of the universe on the proposition *God exists*, is to also confirm the existence of the laws, for without the laws there would be no universe. To then probabilistically confirm the existence of the laws on the proposition *God exists* is to confirm the existence of the universe; there is a distinct lack of probabilistic independence in the process. (See Chapter 2 and Appendix 2) To then add the two resultant posterior probability values in a Bayesian cumulative analysis is to unjustifiably double count, weakening any objectivity in the procedure.

EVALUATORS' CONCLUSION

The evaluators consider that naturalistic explanations have equal probabilistic force on the evidence of the finely tuned laws of nature to that of an intelligent design. Thus the evaluators conclude that $\Pr(e|p \,\&\, k) = \Pr(e|\neg p \,\&\, k)$ where e is the evidence of the finely tuned laws of the universe. However, the evaluators also conclude that the teleological argument cannot be employed in a Bayesian cumulative argument that also includes the Cosmological argument in its evidential set.[42]

Evaluators' Judgement

Both evaluators conclude that the hypothetical syndicate would reach a consensus to eliminate the teleological argument from the evidential set to be considered. This is due to the lack of probabilistic independence from the already considered cosmological argument.

Bayesian Conditionalization

No conditionalization by the hypothetical syndicate members will occur and their priors will remain as atheist: 0.17; agnostic: 0.64; theist: 0.91.

§ SECTION THREE: THE ARGUMENT FROM PROVIDENCE

This is the third and final evidential element in this chapter and is concerned with God's provision of care for His creation or lack of care. Indeed, the problem of evil centers around this apparent lack of care, evidenced by the existence of horrendous evils in terms of human and animal suffering. It was Richard Swinburne (2004) that proposed the argument from providence and applied the concept to the divine provision of *opportunity* for intentional human action that is of considerable consequence to themselves, others, animals, and the world: "a choice of destiny." Swinburne's argument is probabilistic in so far as he seeks to determine the confirmatory force of opportunity-providence on the theistic proposition '*God exists.*'

Confirming the Evidence on the Proposition

The argument begins with a description of the natural hazards of the world and the strategies needed to overcome them, in particular human cooperation. Swinburne argues that the benefits from the dangers, successes, opportunities, failures, suffering, joy, and other human states we have come to know are at a level that is indicative of divine provision. Our nature and environment could have been so much worse resulting in excessive failure, depravity, and suffering, or it could have been more opulent, resulting in mankind languishing in soporific luxury with no effort required to make things and themselves better. Swinburne argues that the state of affairs mankind finds itself in is just right to encourage learning, courage, compassion, discovery, and a range of other worthy attributes and pursuits.[43] Thus, God has instantiated a world where the state of affairs therein ensures the maximum potentiality for persons and animals to develop good attributes that would otherwise be denied to them with a different state of affairs. *Prima facie*, it seems that what Swinburne is presenting is a greater good argument for the problem of evil. However, although the argument relates to the problem of evil, it is not specifically about that problem. Rather is about what is the best world to develop those human and animal attributes deemed to be good.[44]

Counter Argument

The argument is a generalist one in so far as overall the state of affairs is at its best for the development of human and animal virtues per se, but individuals may be swamped or destroyed by this state of affairs. Indeed, persons in a continual state of abject suffering are unlikely to assert that the state of affairs they find themselves in is conducive to the development of human virtues; however, this is an issue for the problem of evil in the next chapter. The main problem with Swinburne's argument is that it is based upon what state of affairs we find ourselves in, rather than an analysis of what the best state of affairs would be for the development of the virtues and then how that compares with reality. It is an almost impossible task and the tendency for the theist will be to default to reality and then justify it which is rather begging the question.

Confirming the Evidence on the Negation of the Proposition

The obvious naturalistic explanation is that of evolution by natural selection. Thus, the success of our species and other species is based upon adaptation over eons to a changing environment within a context of stable natural laws. Thus, it is not surprising that we find ourselves in an overall environment conducive to the success of our species. Likewise, multiverse theory would suggest a vast or even infinite collection of universes exhibiting a huge range of natural laws and material states of affairs. Again it is not surprising that this universe and our world therein is conducive to the success of our and other species, for we would not be here otherwise. Should such a state of affairs change through, for example, an asteroid impact or nuclear war, then possibly the state of affairs will not be conducive to the success of our and other species.

Counter Argument

Notwithstanding the aforementioned doubts concerning the sufficiency of evolution by natural selection as an alternative explanation to intelligent design, evolutionary theory does not answer the fine-tuning argument; that is, divine selection of the ideal laws of nature to enable evolution by natural selection. Moreover, despite multiverse theory providing a possible explanation for the fine-tuning argument, it is based upon mathematical cosmology and is thin on evidential support.

Reply

See above for a response to the evidential challenge related to multiverse theory.

EVALUATORS' CONCLUSION

The argument from providence is matched by the alternative naturalistic arguments. Consequently, the existence of a providential world as described by Swinburne, is equally as liable to exist if God exists or does not exist. Notwithstanding, there will be a tendency for the theist to warm to the argument from providence and for the atheist to warm to the naturalistic alternatives. However, the evaluators both believe that a consensus will be reached whereby the existence of an apparent providential world is equally confirmed if the theistic proposition *God exists* is true or false.

Evaluators' Judgement

1. *Atheist's view*: a medium to low value for $\Pr(e|p \ \& \ k)$ and a medium to high value for $\Pr(e|\neg p \ \& \ k)$
2. *Agnostic's view*: a medium to low value for $\Pr(e|p \ \& \ k)$ and a medium to high value for $\Pr(e|\neg p \ \& \ k)$
3. *Theist's view*: a medium to high value for $\Pr(e|p \ \& \ k)$ and a medium to low value for $\Pr(e|\neg p \ \& \ k)$

Consensus

A consensus between all members is expected, with some compromise for the small differences as follows:

$\Pr(e|p \ \& \ k) = 0.5$
$\Pr(e|\neg p \ \& \ k) = 0.5$
$\Pr(e|p \ \& \ k)/\Pr(e|\neg p \ \& \ k) = \lambda = 1$

Prior Probability

The follow-on priors that apply to this evidential element are as follows, atheist: 0.17; agnostic: 0.64; theist 0.91.

Bayesian Conditionalization

The nature of persons, animals, and our world is clear for all to see, therefore it is reasonable to assume conditionalization by the hypothetical syndicate members.

Posterior Probability

A consensus was reached on a λ-value of 1; consequently, there is no confirmation or disconfirmation of the theistic proposition. Hence, posterior probability values match the prior probability values: atheist: 0.17; agnostic: 0.64; theist 0.91

The above posterior probability values have not diverged from the prior probability values as a consequence of the argument from providence. Thus, these unchanged posterior probability values become the prior probability values for the next evidential element; the problem of evil, which comprises the whole of the next chapter.

NOTES

1. The horse-racing terminology I use in the title of this and other chapters alludes to the Frank Ramsey and Bruno De Finette notion of Coherent betting in the rational application of subjective probability introduced in Chapter 2.

2. Old Evidence; Ch. 2.

3. For example, Swinburne R. (2004); pp. 137–145.

4. Craig W. L. and Smith Q. (1995); Ch. 1.

5. The principle of sufficient reason is primarily associated with Leibniz; there is no fact without there being a sufficient reason for it: Leibniz G. W. (1714) (1973); p. 184; 32.

6. The principle of excluded middle is primarily associated with Aristotle. In sentential logic it states that '*p* or not *p*' is valid and necessarily true on all interpretations of *p*.

7. Braine D. (1998); p. 42.

8. Swinburne R. (2004); p. 53ff.

9. *Ibid*; p. 61ff.

10. *Ibid*; Ch. 7.

11. Howson C. & Urbach P. (1991); p. 291.

12. Howard-Snyder (1996); p. 294.

13. Swinburne R. (2001); pp. 83–99.

14. Swinburne R. (2004); p. 97.

15. *Ibid*; p. 59.

16. Howson C. (2011); p. 126f.

17. Howson C. (2000); p. 206.

18. Gale R. M. and Pruss A. R. (1999); pp. 365–380.

19. If the proposition *God exists* is necessarily true, then the proposition *either God exists or God does not exist* is necessarily true.

20. Mackie J. L. (1982); p. 84.

21. Russell B. and Copleston F. (1964); pp. 168–178 also Hume D. [1779] (1998); II, Hume D. [1748] (1993); IV.

22. Koons R. C. (1997); 8.1.

23. Swinburne R. (2004); p. 76.

24. Kuhn T. S. (2012).

25. Hume D. [1779] (1998); II; p. 15.

26. On the cosmic scale, Hume argues that analogy, being a form of induction, requires many examples to establish a clear causal inference and the universe is limited to one example; moreover, our observation of it is limited to a small segment. Consequently, in Hume's lights, the teleological argument from analogy exhibits a low probability: *Ibid*; XII; p. 88. This, argument is based Hume's understanding of causation; however, limited causal observations of the universe can warrant unobserved conclusions, e.g., quantum field theory.

27. Paley W. [1802] (2008); Chts.1 and III.

28. Swinburne R. (2004); pp. 172–184.

29. *Ibid*; p. 189f.

30. *Ibid*; p. 188f.

31. Lennox J. C. (2009); Chs. 6-10 for an excellent referenced counter argument to macro-evolutionary theory and a persuasive challenge to a sufficient scientific explanation for the emergence of life on Earth.

32. Positive gravity pulls in, whereas negative gravity pushes out.

33. Einstein's limitation on speed; that is, the speed of light is within space, not space itself.

34. Cosmologists have proposed other multiverse possibilities in addition to the inflationary multiverse: landscape multiverse; branes; quantum multiverse; holographic multiverse; simulated multiverse: Greene B. (2011).

35. Interestingly Martin Rees suggest the possibility of a set of multiverse laws with 'bylaws' for each universe: Rees M. (2003); p. 218.

36. Greene B. (2011); Ch. 7.

37. Livio M. (2009).

38. The strong anthropic principle states that the instantiation of the universe ensured the existence of mankind on the grounds that, on the basis of quantum mechanics, existence cannot be predicated to anything unless it is first observed: Barrow J. D. and F. J. Tipler (1996); Ch.7.

39. Sober E. (2009); pp. 76–85.

40. Howson C. & Urbach P. (1991); p. 274.

41. Swinburne R. (2004); 156f.

42. Swinburne takes a different view to these two conclusions: Swinburne R. (2003); p. 120.

43. Cottingham J. (2015); p. 25.

44. Swinburne R. (2004); Ch.10.

Chapter 4

The Bookie's Favorite

The Problem of Evil

This chapter[1] follows from and relates to the previous one in so far as it is concerned with a central evidential element that probabilistically confirms or disconfirms the theistic proposition *God exists*. The same procedure will apply to this evidential element as with the previous evidential elements.

Because of its negative impact on the theistic proposition, the arguments justifying God in the face of evil or refuting His existence because of evil are extensive and much ink has been spilt on the subject; hence, to do justice to the argument I have dedicated a whole chapter to it. Moreover, there is an additional reason for separating it from the previous and following chapters. The previous chapter is concerned with the evidence of divine creativity and the following chapter with evidence from the divine/human relationship. The problem of evil has a foot in both camps; natural evil emerges from the nature and laws of the universe, and moral evil from the intentional malevolence of persons.

The Reader's Assessment

If the reader is assessing their own degree of belief in the theistic proposition *God exists*, then they must bring their final probability value of the theistic proposition on the evidence of the previous chapter forward to this one. This value will then represent their prior probability value of the truth of the theistic proposition; prior, that is, to consideration of the problem of evil.

Heartlessness

The problem of evil has another facet that is not shared by the other evidential elements of Chapters 3 and 5. That facet is the pain and anguish of real suffering that all humans face in their life, some more than others and some so devastating to be life changing. Thus, as Terrence Tilly (2000) argues, to approach the subject in a theoretical way is an evil "…through the radical disjunction of 'academic' philosophical theology from 'pastoral' counsel." [Tilly, 2000, 3] The reader may also feel that such an approach, an approach I take herein, is callous and heartless, and for this I apologise. However, there are two problems of evil, the problem of actual suffering and the problem of reconciling this with the existence of God. They are interrelated, but addressed in different ways, and to entangle them is to do neither justice.[2]

§ SECTION ONE: NATURAL AND MORAL EVIL

The problem of evil is the most forceful disconfirming evidential element to be considered. Indeed, the problem is ably expressed by David Hume [1779] (1998) when considering the abilities of God through "Epicurus' old questions…Is he willing to prevent evil, but not able? then he is impotent. Is he able but not willing? then he is malevolent. Is he both able and willing? whence then is evil?" [Hume, [1779], 1998, Part X; 63][3]

There are deductive, inductive, and abductive arguments that challenge the very existence of God based on the evil manifested in the world. Evil in this argument does not refer directly to any supernatural force, such as demonic or satanic influence, although the possibility of supernatural evil should not be dismissed. Rather, it refers to human and animal suffering caused by intentional human action, termed *moral evil* or by natural causes; for example, accidents, animal attack, disease, natural disasters, and so on, termed *natural evil.*

The probabilistic disconfirming force of evil on the proposition *God exists* results from, *inter alia*, three of the essential characteristics of God; that is, omnipotence, omniscience, and omni-benevolence. If God is all powerful, knows all, and has maximal love for His creation why is there so much terrible evil in the world; *prima facie*, it is because He does not exist.

Abbreviations

There are some abbreviations employed in the following arguments: 'omni-being' means an omnipotent, omniscient, and omni-benevolent being and the 'omni-predicates' refer to the predicates 'is omnipotent,' 'is omniscient,' and 'is omni-benevolent.'

Confirming or Disconfirming the Evidence on the Proposition

Like other evidential arguments, the problem of evil is normally presented as the force of the evidence on the proposition; thus, evil disconfirms the existence of God. In Bayesian terms, it takes the form of the posterior probability

$Pr(p|e \& k)$. However, I am taking the likelihood ratio $Pr(e|p \& k)/Pr(e|\neg p \& k)$ as the confirming and disconfirming procedure with understanding of the problem of evil arguments residing in the background knowledge *k*.

Defence

The problem of evil is fundamentally an atheistic argument in support of the negation of the theistic hypothesis; *God does not exist*. Hence, assuming the existence of an omni-being, we would expect less moral and natural evil in the world and assuming the non-existence of an omni-being we would expect more moral and natural evil in the world. Therefore, the theist has traditionally taken a defensive position, providing counter arguments to those put forward by those apologists supporting this disconfirming argument.

One such defence is the greater good defence based upon the assertion that God, as the omni-being, permits moral and natural evil to exist in the world because a good, more valuable than the evil, comes about, or if He were to eliminate it, an evil worse than that eliminated would result.

Theodicy

There are confirming arguments for $Pr(e|p \& k)$; theodicies provide that viewpoint in the argument. Theodicies differ from defences in that they attempt to provide a complete explanation for the existence of evil in the world given the existence of God and I introduce some below in Section Two of this chapter. To be expected, there is considerable overlap between theodicies and defences.

The Deductive Argument

This conventionally takes the following form:

1. *Premise*: if God exists He is an omni-being
2. *Premise*: an omni-being eliminates all evil
3. *Premise*: evil exists
4. *From 1, 2 & 3*: if God and evil exist then God does not eliminate all evil
5. *From 3 & 4*: God does not exist

Counter Argument

This is a very strong argument and renders $Pr(\neg p|e \& k)$ certain; a probability value of 1 with even background knowledge *k* irrelevant. However, the deduction depends on two propositions:

1. God is an omni-being
2. God eliminates all evil

The denial of any one of the omni-predicates would render God's existence compatible with the existence of evil in the world. However, I argue in Chapter 1 that God is indeed an omni-being; all the omni-predicates are essential attributes of God. However, the proposition God eliminates all evil because of the omni-predicates is questionable because it presumes questionable auxiliary assumptions; for example, God does accept greater good scenarios. (See Appendix 1) An obvious defeating example would be that of a medical operation that can cause much suffering, but the outcome of a successful operation can outweigh such suffering: an overall benefit for the patient. Therefore, a more focused and specific rendering of 2 and 3 in the deduction is called for.

Gratuitous Evil

Let 'Bambi' and 'Sue' be cases of horrendous natural and moral evil respectively. 'Bambi' referring to an imaginary fawn who is caught in a forest fire, is badly burnt, and dies an excruciating death over a long period, unseen by a person. Although not actual, it is a scenario that has probably occurred many times. 'Sue' referring to an actual case of a mother's boyfriend raping and murdering her young daughter.[4] These two cases represent horrendous natural and moral evil where:

1. No greater good can be conceived of for God, as the omni-being, permitting them to occur and,
2. No greater evil can be conceived of that would emerge if God eliminated these specific evils.

The 'no greater good if the evil is ignored' and 'no greater evil if eliminated' is what is meant by *gratuitous evil* and 'Bambi' and 'Sue' are cited as paradigms of it. The deductive argument then proceeds thus:

1. *Premise*: if God exists He is an omni-being
2. *Premise*: an omni-being eliminates all gratuitous evil
3. *Premise*: 'Bambi' and 'Sue' are cases of gratuitous evil and occurred
4. *From 1 & 2*: if God exists and 'Bambi' and 'Sue' occurred then God does not eliminate gratuitous evil
5. *From 3 & 4*: God does not exist

Counter Argument

The concept of gratuitous evil applied to specific cases such as 'Bambi' and 'Sue,' is open to challenge in its use within a sound deductive argument. There may be a greater good beyond our ken, but known unto God. Thus, the deduction still fails on this gratuitous evil refinement. The best that can be hoped for with premise 3 is that *probably* 'Bambi' and 'Sue' are cases of gratuitous evil, culminating in the conclusion 'probably God does not exist.'

The Inductive Argument

One inductive argument from evil to the negation of the proposition *God exists* is based upon a deontological moral perspective.[5] The argument is an inductive generalization from *known* moral properties of a state of affairs, to a *complete* moral perspective of the state of affairs that include known and unknown moral properties. Consider the following:

1. Assuming God exists, then God's inaction in eliminating the apparent gratuitous evil of 'Bambi' and 'Sue' possesses *known* wrong-making-properties and *known* right-making-properties (that is known by us rather than God). Given they are cases of gratuitous evil, they do not possess any *known* right-making-properties. Thus, the known wrong-making-properties completely *outweigh* the known right-making-properties.
2. Let there be a *total* of wrong- and right-making properties for God's inaction in the cases of 'Bambi' and 'Sue.' This total comprises both known and unknown wrong- and right-making properties (that is known and unknown by us).
3. The inductive move is from the known to the total. Given that the known wrong-making-properties outweigh the known right-making-properties, then the total wrong-making-properties outweigh the known right-making-properties.

The earlier gratuitous evil deductive argument faltered because of the possibility of goods beyond our ken, but known unto God. However, by inductively moving from *known*-wrong-making properties (deontological parlance) to *total*-wrong- making properties; both outweighing known right-making-properties, the deductive argument converts to an inductive one and holds.

Counter Argument

Assuming the validity of the deontological moral perspective, the argument hinges on the inductive step:

1. All observed things of a type have certain properties; therefore, *all* things of that type have those properties.

The generality of the use of *all* is a demanding inductive argument, whereas the following is less demanding:

2. All observed things of a type have certain properties; therefore, the *next thing observed* of that type will have those properties.

It is a question of probability; *all* is all-encompassing, a universal generalization and one anomaly would refute it, whereas the *next thing observed* based upon regularity, could easily miss the anomaly. Thus, the *all*-factor based upon the *some*-factor is more liable to fail than the *next*-factor based upon the *some*-factor. However, the above argument requires the *all*-factor for it to hold, as the mere possibility of an anomaly would destroy the argument. Moreover, the *some*-to-*all* argument is nearer to a law than induction through regularity; however, the inductive element of the argument is not a law or even the result of a natural law.[6] Moreover, as Hume ably argued, such an inductive move is circular.[7]

Draper's Inductive Argument

Paul Draper (1996) provides an interesting inductive argument. Let…

q *be* "…neither the nature nor the condition of sentient beings on earth is the result of benevolent or malevolent actions performed by nonhuman persons." [Draper, 1996, 13][8]
p be the theistic proposition *God exists*

Less evil in world is expected given p than q, particularly gratuitous evil, as Draper's q implies a naturalistic context that is amoral, non-sentient, and certainly not omnipotent. With less detail and a slightly different application of probability theory than Draper's, the argument proceeds as follows:

1. *Premise*: q entails $\neg p$
   That is, $\neg p$ cannot be false and q true
2. *Premise*: $\Pr(e|q) > \Pr(e|p)$
   That is, evil is more likely to exist given q than p
3. *Premise*: $\Pr(q) \geq \Pr(p)$
4. *Conclusion*: $\Pr(\neg p|e) > \Pr(p|e)$

This can be demonstrated from Bayes' theorem:

5. *From Bayes' theorem*:

$$\Pr(q|e) = \frac{\Pr(e|q)\Pr(q)}{\Pr(e)}$$

6. *From Bayes' theorem*:

$$\Pr(p|e) = \frac{\Pr(e|p)\Pr(p)}{\Pr(e)}$$

7. *From 5 & 6*:

$$\frac{\Pr(q|e)}{\Pr(p|e)} = \frac{\Pr(e|q)}{\Pr(e|p)} \times \frac{\Pr(q)}{\Pr(p)}$$

8. *From 2, 3 & 7*: $\Pr(q|e) > \Pr(p|e)$
9. *From 1 & 8*: $\Pr(\neg p|e) > \Pr(p|e)$
10. *From axiom 2*: $\Pr(\neg p|e) + \Pr(p|e) = 1$
11. *From 9 & 10*: $\Pr(p|e) < 0.5$ and $\Pr(\neg p|e) > 0.5$

From the above, the existence of evil *e* in the world, particularly gratuitous evil, probabilistically favors $\neg p$ over *p*.[9]

Counter Arguments

Premise 1; that is, *q* is logically incompatible with *p* is also a concern as it seems that *q* and *p* could be compatible if God withdrew His presence in the universe post-creation having established the laws of nature that ensure the emergence of life. (See *The Theodicy of Divine Withdrawal* herein below). However, a different *q* would account for this objection as Draper suggests: "There exists an omnipotent and omniscient person who created the Universe and who has no intrinsic concern about the pain or pleasure of other beings." [Draper, 1996, 13]

The conclusion that $\Pr(p|e) < 0.5$ is somewhat limiting; $\Pr(p|e) = 0.499$ would suffice, but intuitively many would think it should be much lower, particularly given the existence of gratuitous evil.

The absence of background knowledge *k* from the above, in which additional confirming and disconfirming evidence would be included, impacts on the Bayesian process as a stand-alone calculation. To undertake a Bayesian analysis of the problem of evil without accounting for all other known relevant evidence is probabilistically unsound and self-deceptive. Evidence such as religious experience, the existence of the universe, the finely tuned laws of nature, miracles, and more should either be included in the background knowledge *k* or accounted for in a cumulative process like Swinburne's or as undertaken herein.

Rowe's Bayesian Analysis

William Rowe (1996)[B] employs logical Bayesianism in an analysis of the problem of evil using 'Bambi' and 'Sue' as the evidential paradigm of gratuitous evil. Using Rowe's symbols let…

G be the proposition 'an omni-being exists.' (G being the equivalent of *p* that I use throughout)
P be the gratuitous evil statement "no good *we know of* justifies an omni-being permitting 'Bambi' and 'Sue.'"
Q be an additional gratuitous evil statement 'no good *at all* justifies an omni-being permitting 'Bambi' and 'Sue.''"[10]

The argument proceeds as follows:

1. *Premise*: $\Pr(G|k) = 0.5$

This value appears to be based upon the principle of indifference [Rowe W. (1996)[B]; p. 265f & p. 283n10]

2. *Premise*: $\Pr(P|G \,\&\, k) = 0.5$

This also appears to be an allocation based upon the principle of indifference, although Rowe acknowledges that so long as its value is less than 1 it has minimal impact on the result. [Rowe W. (1996)[B]; p.268]

3. *Premise*: $\Pr(P|\neg G \,\&\, k) = 1$

P states "no good *we know of* justifies an omni-being permitting 'Bambi' and 'Sue.'" ¬G states that an omni-being does not exist, so P is certain on the grounds that we cannot know of a good that justifies a non-existent being. [Rowe W. (1996)[B]; pp. 265 & 268]

4. *Premise*: $\Pr(P|Q \,\&\, k) = 1$

Q conceptually entails P; 'no good at all justifies…' clearly entails '…no good we know of justifies…'. If something is false we cannot know it as true. [Rowe W. (1996)[B]; p. 267]

5. *Premise*: $\Pr(Q|G \,\&\, k) = 0$

This is a case of metaphysical rather than logical necessity. If an omni-being exists there must be a justification for 'Bambi' and 'Sue'. This is true because of the essential nature of the concepts employed; the omni-predicates ensure the falsity of Q.

6. *Premise*: $Pr(Q|\neg G \& k) = 1$

This is a case of logical entailment, i.e., it is certain that no good can justify an omni-being if such a being does not exist.

7. *From the theorem of total probability; 1; 2; 3*:
   $Pr(P|k) = Pr(P|G \& k)\, Pr(G|k) + Pr(P|\neg G \& k)\, Pr(\neg G|k)$
   $Pr(P|k) = (0.5 \times 0.5) + (1 \times 0.5) = 0.75$
8. *From the theorem of total probability; 1, 5, 6*:
   $Pr(Q|k) = Pr(Q|G \& k)\, Pr(G|k) + Pr(Q|\neg G \& k)\, Pr(\neg G|k)$
   $Pr(Q|k) = (0 \times 0.5) + (1 \times 0.5) = 0.5$
9. *From Bayes' theorem; 4; 7; 8*:
   $Pr(Q|P \& k) = Pr(P|Q \& k)Pr(Q|k)/Pr(P|k)$
   $Pr(Q|P \& k) = (1 \times 0.5)/0.75 = 0.67$
10. *Premise:* Q metaphysically entails $\neg G$ (an omni-being cannot exist if no good justifies 'Bambi' and 'Sue' because of the breach of omni-benevolence)
11. *Premise*: $\neg G$ entails Q (if an omni-being does not exist no good can justify an omni-being)
12. *From 10 & 11*: $Pr(Q|P \& k) = Pr(\neg G|P \& k)$
13. *From 12 & axiom 2*: $Pr(G|P \& k) = 1 - Pr(Q|P \& k)$
14. *From 9 & 13*: $Pr(G|P \& k) = 1 - 0.67 = 0.33$

Thus, Rowe concludes:

> ...P not only makes Q more likely than it otherwise would be but also makes it more likely than not...The argument proceeds inductively from P to Q and deductively from Q to ~G...So it does seem that we are justified in holding not only that P makes G less likely than it would otherwise be, but also justified in holding that P makes G less likely than ~G. [Rowe W. (1996)[B]; p.270] (Rowe uses the symbol ~ for 'not')

Counter Arguments

Based upon a desire not to beg the question, Rowe applies the principle of indifference to value $Pr(G|k)$ as 0.5;[11] that is, given alternative, exhaustive, and exclusive functions, in this case G and $\neg G$. Rowe then divides the probability continuum by half, allocating 0.5 to G and $\neg G$ respectively.[12] However, Rowe believes that $Pr(G|k) < 0.5$ on the basis of "the abundance of various evils in the world" [Rowe, 1996[B], 288n10], which is a gesture to subjective probability. If this is the case, then we would expect a variety of prior probability values for Pr(G|*k*) differing between the atheist, the agnostic, and the theist. It will be clear to the reader that this is how I believe the argument should proceed, preferably through the process of intersubjective probability. However, Rowe settles on Pr(G|*k*) = 0.5 on, I believe, the principle of indifference; a questionable procedure in logical probability given the paradoxes associated with the principle.[13] Rowe suggests his choice of 0.5 is in order not to beg any questions, which amounts to the same. Outside of subjective probability, there is no sound principle to allocate a prior value and Rowe's selection of 0.5 is arbitrary.

Michael Tooley (2015) suggests that Pr(G|*k*) < 0.5 is a reasonable assessment given that an omni-being is but one member of a family of properties including possible individuals that are omnipotent and omniscient, but vary from being a perfectly good being (omni-benevolence in my terminology) to a perfectly evil being and a perfectly morally indifferent being.[14] However, it could be argued that variations from perfect can be included in the moral set; for example, almost perfectly morally good, almost perfectly morally evil and almost perfectly morally indifferent. Given these degrees of good and bad, then there are infinite possibilities within the set. If any one of this infinite set of individuals is no less liable to be instantiated than another; for example, than the omni-being, then this would drive Pr(G|*k*) to zero on the principle of indifference and Rowe's argument would fail as Bayes' theory cannot proceed with Pr(G|*k*) = 0. However, in Chapter 1 I argue that omni-benevolence directly follows from the divine attributes of omnipotence and omniscience so such variations to omni-benevolence are, in my view, not applicable.

Rowe has not accounted for total evidence in *k*. This does not directly refute his conclusion; he could extend his analysis to account for all the evidence in a cumulative analysis as Swinburne (2004) has done. However, Rowe does not include the other essential evidential elements that Swinburne addresses, only background evidence relevant to the problem of evil in his argument.[15] Unless a cumulative Bayesian analysis is undertaken, the posterior function Pr(G|P & *k*) should account for more than just the probability of G on P and background knowledge related only to the problem of evil, but P and background knowledge of all the relevant evidence that confirms or disconfirms G.[16]

There is a further problem with Rowe's argument; consider the following two propositions:

P: *No good we know of justifies an omni-being permitting 'Bambi' and 'Sue'*
Q: *No good at all justifies an omni-being permitting 'Bambi' and 'Sue'*

In the case of P, the proposition asserts that, from our background knowledge, we cannot think of anything that justifies an omni-being permitting 'Bambi' and 'Sue.' In the case of Q, the proposition states that there is no good that provides such a justification. It appears to be an inductive move from our lack of knowledge to an expression of certainty; that is, there is a probabilistic relationship where P confirms Q.

It seems contrived to introduce a claim that the proposition *an omni-being does not exist* (¬G) entails P as Rowe's argument does in premise 3. P is concerned with a lack of justifying knowledge and to claim certainty on the basis of assuming ¬G in the probability function Pr(P|¬G & *k*) is to conceal the main purpose of the argument; that is, an inductive move from a lack of knowledge to a degree of certainty. To avoid such concealment and still keep the likelihood function Pr(P|¬G & *k*) it should change to Pr(P*|¬G & *k*) where P* states *no good we know of justifies an omni-being (if one exists) permitting 'Bambi' and 'Sue.'*[17] Q should also change to Q* on the same principle to *no good at all justifies an omni-being (if one exists) permitting 'Bambi' and 'Sue.'* These changes bring the argument back on track to its obvious purpose where the proposition ¬G has confirming or disconfirming force on P*.

Returning to Rowe's logical probability and assuming the subjunctive clauses in P* and Q*, ¬G does not entail Q* so the move from 10 and 11 in Rowe's argument as presented, to Pr(Q*|P* & *k*) = Pr(¬G|P* & *k*) in 12 is not justified and consequently Pr(G|P* & *k*) = 0.33 in 14 would not follow. Moreover, given the possible application of theodicies in Pr(P*|G & *k*) Rowe's principle of indifference value of 0.5 in 2 is too high and consequently, so is Pr(P*|*k*) in the total probability calculation in 7. Rowe's high value for Pr(Q|P & *k*) is not surprising as P clearly confirms Q and would as P* and Q*.

Rowe Drops Q

Later, Rowe bypasses the necessity for Q and proceeds directly from P to ¬G.[18] Assuming premises 1, 2, 3, and 7 from Rowe's analysis above, a Bayesian analysis determines Pr(G|P & *k*) as follows:

a. *From the theorem of total probability & premises 1, 2 & 3*:
   Pr(P|*k*) = Pr(P|G & *k*)Pr(G|*k*) + Pr(P|¬G & *k*)Pr(¬G|*k*) = (0.5 × 0.5) + (1 × 0.5) = 0.75
b. *From Bayes' theorem, a & 7*:
   Pr(G|P & *k*) = Pr(P|G & *k*)Pr(G|*k*)/Pr(P|*k*) = 0.5 × 0.5/0.75 = 0.33
c. Pr(G|P & *k*) > Pr(G|*k*) hence disconfirmation.

No reference to Q is required, but the function Pr(P|¬G & *k*) = 1 on the basis entailment is employed again in a.

Plantinga's Response to Rowe

Alvin Plantinga (1998), reflecting upon Rowe's new argument, compares Rowe's 'atheistic' P with a 'theistic' one as follows:

> P^A: "Neither E1 nor E2 is such that we know that no known good justifies a perfect being in permitting it" [Plantinga (1998); p. 536][19] [E1 and E2 being Bambi and Sue respectively)
> "P: No good we know of justifies God in permitting E1 and E2" [Rowe, 1996^B, 270]

He argues that G entails P^A [20] just as ¬G entails P.

Plantinga's argument proceeds as follows: Let...

1. Pr(P^A|¬G & *k*) = 0.5 and Pr(¬G|*k*) = 0.5 (As highly speculative as Rowe's Pr(P|G & *k*) and Pr(G|*k*))[21]
2. *From the theorem of total probability*:
   Pr(P^A|*k*) = Pr(P^A|G & *k*)Pr(G|*k*) + Pr(P^A|¬G & *k*)Pr(¬G|*k*)
   Pr(P^A|*k*) = (1 × 0.5) + (0.5 × 0.5) = 0.75
3. *From Bayes' theorem & 1 & 2*: Pr(¬G|P^A & *k*) = Pr(P^A|¬G & *k*)Pr(¬G|*k*)/ Pr(P^A|*k*)
   Pr(¬G|P^A & *k*) = (0.5 × 0.5)/0.75 = 0.33 and therefore Pr(G|P^A & *k*) = 0.67

This is an identical result to Rowe's Pr(G|P & *k*) = 0.33 and hence Pr(¬G|P & *k*) = 0.67

As ¬G entails P, P confirms ¬G for Rowe[22] and likewise, for Plantinga, as G entails P^A then P^A confirms G. Moreover, given that identical results are obtained by employing equivalent values in each case, then Rowe's atheistic argument from P seems counterbalanced by Plantinga's theistic argument from P^A. Indeed, taking total evidence into account; that is, the theist's and atheist's interpretation of God justifying goods for E1 and E2 (P and P^A), then Pr(G|P & P^A & *k*) = 0.5; that is, equal to Pr(G|*k*); hence, no confirmation.[23]

Likelihood Ratio

Returning to the subjunctive form of Rowe's P; namely P*: *no good we know of justifies an omni-being (if one exists) permitting 'Bambi' and 'Sue,'* in the likelihood functions Pr(P*|G & *k*) and Pr(P*|¬G & *k*). In the former function, the attributes of the omni-being (G) ensure there is a justifiable reason for the omni-being permitting 'Bambi' and 'Sue'; an omnipotent, omniscient, and omni-benevolent being could not exist if there were no justifying good.[24]

Let J be a good that justifies the omni-being permitting 'Bambi' and 'Sue,' then, Pr(P*|G & *k*) is disconfirming as there must be a J which reduces the probability of P*. With Pr(P*|¬G & *k*) there may still be a J, despite the non-existence of the omni-being, that would justify the omni-being if one did exist; therefore, Pr(P*|¬G & *k*) is also disconfirming. However, they are not equally disconfirming as the necessity of J in Pr(P*|G & *k*) has more disconfirming force upon P* than the possibility of J in Pr(P*|¬G & *k*).

By way of an analogy, consider the proposition *Elizabeth I is Queen of England*. Also, the proposition *a queen sits on the English throne* and the proposition *Edmond does not know a queen sits on the English throne*. If the proposition *Elizabeth I is Queen of England* is true, then so is the proposition *a queen sits on the English throne*. However, if Elizabeth I does not exist, it is still possible that there is a queen that sits on the English throne, just not her. *Ceteris paribus*, the probability that *Edmond does not know that a queen sits on the English throne* is greater if Elizabeth I does not exist than if she does.

From the above, Pr(P*|G & *k*) < Pr(P*|¬G & *k*) and consequently, the likelihood ratio Pr(P*|G & *k*)/Pr(P*|¬G & *k*) would be less than 1. From Bayes' theorem, in its likelihood ratio form; that is:

$$\Pr(G|P^* \;\&\; k) = \frac{\lambda \times \Pr(G|k)}{[\lambda \times \Pr(G|k)] + [1 - \Pr(G|k)]}$$

Where λ is the likelihood ratio Pr(P*|G & *k*)/Pr(P*|¬G & *k*). Given that this ratio is less than 1, P* & *k* would disconfirm G; that is, Pr(G|P* & *k*) < Pr(G|*k*).

Bayesian conditionalization requires the evidence to be true; namely, P*: *no good we know of justifies an omni-being (if one exists) permitting 'Bambi' and 'Sue.'* Given the contestation surrounding the problem of evil, it is evident that P* is true. This is not to claim that there is no good; indeed, if God exists there must be, it is just that we do not *know* what it is. There are disputed suggestions; for example, the free-will defence, but a claim of knowledge is *inter alia*, a claim of truth and, without assuming God existence, we cannot be certain of this. Therefore, I submit that whether one is an atheist or a theist, on this matter Bayesian conditionalization is justified.

If Pr(G|*k*) = 0.5 as Rowe suggests, then Pr(G|P* & *k*) < 0.5 and P* disconfirms the existence of God. However, as stated above, that is to ignore other relevant evidential elements addressed in Chapters 3 and 5. Moreover, a theist would have a prior probability value of Pr(G|*k*) far greater than 0.5, which is an agnostic's prior degree of belief. The atheist would have a prior degree of belief much less than 0.5. This highlights the importance of total evidence and the advantage, in my view, of subjective probability over logical probability in this context.

The Moore Shift Defence

William Rowe (1996)[A] elicits what he terms the G. E. Moore shift.[25] Rowe suggests that the shift presents the theist with a possible challenge to the argument from evil. The shift can be expressed as: $(g \,\&\, q \rightarrow r) \equiv (\neg r \,\&\, q \rightarrow \neg g)$ and selection between the two equivalents as the soundest argument, depends upon the truth of g or $\neg r$. Thus, assuming gratuitous evil as so defined above, the argument proceeds as follows:

Let the atheistic argument be:
If g (gratuitous evil exists) *and*
q (an omni-being would eliminate gratuitous evil) *then*
r (an omni-being does not exist)
From the G. E. Moore shift, let the theistic argument be:
If $\neg r$ (an omni-being exists) *and*
q (an omni-being would eliminate gratuitous evil) *then*
$\neg g$ (gratuitous evil does not exist)

The arguments are balanced in validity and the theist would argue that there are grounds for accepting $\neg r$ and that $\neg g$ follows, whereas the atheist would argue that there are grounds for accepting g and that r follows. Thus, much depends upon whether the theist has strong grounds for accepting $\neg r$ as true and being able to explain $\neg g$. The grounds for accepting $\neg r$ will emerge from the Bayesian analysis of Chapters 3 to 5 and an explanation for $\neg g$ must be addressed in this chapter.

The 'Noseeum' Defence

Perhaps what we as persons see as gratuitous evil is not seen as gratuitous by God. That is, there is a greater good resulting from cases of gratuitous evil that is beyond our ken, but known unto God. To this end, Stephen Wykstra (1990) introduces the Condition of Reasonable Epistemic Access (CORNEA):

> On the basis of cognized situation *s*, human H is entitled to claim 'It appears that p' only if it is reasonable for H to believe that, given her cognitive faculties and the use she has made of them, if p were not the case, s would likely be different than it is in some way discernible by her. [Wykstra, 1990, 152]

From CORNEA it can be asserted that a person is warranted to claim that *there is no p* on the basis of *it seems there is no p* if that person would be expected to perceive *p* or discover *p* or comprehend *p* (or any other relevant cognitive functions) if *p* were there; in short if the person possesses the wherewithal to perceive *p*. (note that Wykstra's use of the symbol *p* does not represent the *p* used herein as the symbol for theistic proposition *God exists*)

Wykstra in his 1996 paper termed assertions like *there is no p* above, a 'noseeum' inference,[26] for if there are goods that justify an omni-being permitting gratuitous evil like 'Bambi' and 'Sue' then it is reasonable to assume we would 'seeum.' Wykstra's counter argument is that, given God's omniscience, it is possible that such greater goods are beyond our ken. Humans do not possess the cognitive wherewithal in comparison to an omniscient being to be certain that there are no greater goods. Wykstra employs an adult-baby cognitive levels analogy to the human-God comparison on this matter.[27] However, the loving compassion shown by a parent to their suffering child seems absent with God and God can intercede unlike a parent in many cases; the analogy is questionable.[28] In short, Wykstra's position is that things would not strike us differently with respect to gratuitous evil given the existence rather than the non-existence of an omni-being.[29]

Counter Argument

Richard Gale (1996) argues that doubting the assertion 'noseeum-therefore-none' based on the 'beyond our ken' argument leads to radical scepticism. "A parallel argument can be constructed to show that every proposition is such that we humans could not be epistemically justified in making any claim, probabilistic or otherwise, about its truth value." [Gale, 1996, 208] He asks us to consider the following proposition:

1. An omnipotent, omniscient, and *malevolent* being exists (*d*).
2. *d* refers to a demonic supernatural being, making the truth of *d* beyond our ken.
3. Let there be a negative existential proposition 'there is not a lion in the room' with noseeum evidence confirming its truth.

Because of the demon's omni-malevolence, if it is believed that it intentionally fools us with respect to the evidence, we would probably doubt the confirming force of the noseeum evidence and assert 'there is a lion in the room even though we can't see it.'

Reply

However, we do intuitively differentiate between reasonable and unreasonable doubts; for example, doubting there is no extra-terrestrial life compared to doubting that the Earth is more than 100 years old. Thus, doubts about 'noseeum-therefore-none' are more like the reasonable type.[30]

Things Could Be Much Worse

It could be argued that God does in fact eliminate gratuitous evil and that what we observe is merely apparent gratuitous evil. This is a *things could be much worse* argument whereby human and animal suffering does result in a greater good, or that its elimination would lead to worse suffering. The inability to appreciate the non-gratuitous nature of the suffering is a consequence of human ignorance compared to that of an omniscient being. Moreover, if God revealed its non-gratuitous nature, then there would be a disinclination for people to eliminate it because of its revealed greater good. This could lead to a loss of benefit of a different nature; that is, the fortitude gained from attempts to overcome the cause of the suffering along with the concomitant development of human knowledge, understanding, and skills.[31] Thus, it is better for God to keep mankind in the dark with respect to its non-gratuitous nature.

Counter Argument

Taking-up the *things could be much worse* argument, William Hasker (1992) distinguishes between two types of gratuitous evil:

1. *Genuine gratuitous evil*; an evil the elimination of which would not also eliminate a greater good and
2. *Apparent gratuitous evil* the elimination of which would result in the loss of
   (a) a non-moral greater good or
   (b) the greater good of moral fortitude; for example, "…knowledge, prudence, courage, foresight, cooperation, and compassion…" [Hasker, 1992, 39]

If it were known that God eliminates the genuine form and leaves the apparent form, there would be a disinclination for persons to eliminate it because of the known loss of the greater good of moral fortitude together with the non-moral greater good. However, such disinclination would itself undermine moral fortitude by not engaging with ameliorating human suffering. If, on the other hand, it was not known that God eliminated the genuine form, then persons would probably believe the apparent form was genuine and would attempt to eliminate it along with its unknown concomitant greater good. Such elimination, if successful, would create a need for a divine top-up of the apparent form to maintain greater-good levels. This divine topping-up process would exacerbate the genuine/apparent illusion and as such God would be deceiving persons, which would be incongruous with omni-benevolence. Thus, to avoid such deception and at the same time provide for the greater good, both the genuine and apparent forms would need to be provided for by God. Thus, argues Hasker the existence of God and genuine gratuitous evil are not logically inconsistent.[32]

Reply

David O'Connor (1998) argues that lack of revelation is not false revelation, thus Hasker's charge of deception may not apply.[33]

The act of ameliorating apparent gratuitous evil is a greater good that could equal the loss of the greater good from the elimination of the apparent gratuitous evil. This lost greater good is the same as that obtained if the apparent gratuitous evil were ignored. However, the elimination of a greater good is genuinely gratuitous and, therefore, as the genuine form cannot occur, then the attempted elimination would fail. Therefore, attempting to eliminate would increase the overall greater good from the attempt plus the subsequent greater good resulting from the failure. This is not equal to the ignoring scenario and therefore there is no need for a divine top-up of the lost apparent gratuitous evil, as the greater good has increased. Therefore, there is no need for deception and therefore no need for the genuine form and, therefore, the existence of God and genuine gratuitous evil is not shown to be compatible by this argument.[34]

§ SECTION TWO: THE FREE-WILL DEFENCE

The most common greater good defence, and theodicy, of moral evil is that of free-will. The form of the argument is that the free-will of persons is so valuable to them and God, that it is worth God tolerating the misuse of free-will to generate moral evil.

Is Free-Will Possible?

To answer this question positively, we must show that persons make, exercise, and are responsible for their unencumbered decisions; decisions that are conscious mental states that result in action for which they can be held responsible.[35]

A materialistic concept of mind may conceive of mental states as just brain states and nothing else.[36] Given this, then the idea of free-will is very questionable, possibly an illusion.[37] What appears to be a conscious decision is an irrelevant sideshow alongside a material causal nexus. This hard-materialistic view, sometimes referred to as strong determinism,[38] is intuitively wrong, as people feel they are masters of their own destiny; their decisions are unencumbered conscious choices. However, the questionability of substance dualism[39] leads us to the inevitable conclusion that mental states are inextricably connected to brain states: a mental state cannot occur without a concomitant brain state occurring. Mental states being more accurately described as *emergent properties* of brain states.[40] Such emergent mental properties supervene on brain states in so far as a mental property cannot occur without a brain state and cannot change unless the brain state does.[41] Thus, a different mental property will be accompanied by a different brain state. By contrast a brain state can occur without a concomitant mental property supervening on it.[42]

Donald Davidson's (1970) Anomalous Monism argument states that all causal events are governed by strict nomological laws and that mental events are not so governed; thus, free-will decisions must be materialistic to be causal.[43] Therefore, if I decide to raise my arm and my arm rises, then brain states engage in the entire procedure and the process of deciding is essentially a material process exhibiting material properties and mental properties, but it is the material properties that are causally efficacious, not the mental properties.[44] Nevertheless, in this arm raising scenario mental properties remain, they are an inevitable consequence of the brain properties. Thus, material properties partner mental properties, somewhat like a great painting; one property being the paint, the other being the art; remove either and it ceases to be a great painting. Although paint and aesthetics can be discussed separately, neither one can be ignored in a holistic vision.[45]

In all appearances, it is the mental property that has causal efficacy; a folk psychology view of free-will. However, this is due to the supervening mental property being almost synchronous with the brain property,[46] giving the impression of mental causation.[47] Notwithstanding, the mental property element of this *mental-material-union* is not a separate inconsequential element despite the material property being the causal realiser; it is the phenomenal element of the union.[48]

Despite this phenomenal element, the act of deciding involves a cause-effect chain of events, governed by strict nomological laws, with a past, present, and future linkage within a causal nexus. Hence, some physical event was sufficient to trigger the arm raising; that is, a prior cause that itself was caused by another physical event that was sufficient to trigger it and so on. This causal exclusivity is known as the principle of the causal closure of the physical; theoretically a full causal explanation can be given without reference to factors outside the causal nexus.[49]

Given this physical determinism whence free-will? I cannot freely choose not to do something if it is determined by a causal nexus; this is the argument of the incompatibilist,[50] although neuroscientific determinism is not universally accepted.[51]

Regarding the *mental-material-union*, perhaps free-will can be envisaged as such a union comprising the mental component *will* and material component *free*. *Will* or *choosing* being the content or quale of a conscious experience. The quale of *will* or *choosing*, supervenes on a concomitant material brain event that can be *free* from causal influence for a period in the relevant causal time-line. Being free in this sense is being unencumbered by another intrinsic cause; for example, another over-riding brain-state or another extrinsic cause; such as, being knocked unconscious. Seen from this holistic perspective of a mental-material union, free-will or freely choosing appear to be appropriate descriptors, although different to the conventional folk perspective. This is a compatibilist notion of free-will not a libertarian one,[52] the latter being freedom of will from any material determinism.[53]

One common criterion for genuine free-will is the modal conditional: '*if p did x he could have done otherwise*' is true if and only if *p* has free-will.[54] On the above revisionist mental-material union model proposed; that is, freely (material) choosing (mental) to do x, there is brain state that could have been different; for example, c-fibres caused to fire instead of d-fibres caused to fire.[55] With c-fibres firing, the mental state of choosing y would emerge. In deterministic material terms, possibility holds; it could have been c-fibres instead of d-fibres if initial causal conditions were different. Hence, in mental state terms, possibility holds; choosing y instead of choosing x. Thus, viewed holistically as a *mental-material union, p could have done otherwise*.

This revisionist conception of free-will seems a little contrived, perhaps it is the power of the person to initiate decisions there and then that counts as free-will; a sense of immediate agency rather than the mere possibility that the initial conditions of the universal causal nexus were different. However, given my argument above, this cannot be an appeal to mental state causation or *causa sui*.[56] Instead, it must be the *mental-material union* that initiates the decision.

The neural structure of the brain is such that a flurry of electrochemical activity can be triggered by a simple material cause, whereby tens of thousands of neurons are suddenly activated in a micro-second, working in a vast variety of combinations and permutations to affect a result greater than the sum of the parts: synergy. That is, the firing neurons are not individually sufficient to generate consciousness, but collectively a conscious state emerges. A state that can be simple, such as a perceptual experience; for example, a smell, or more complex, such as the smell plus a reflection, plus a decision, plus a movement, and the more complex, the greater the combinations and permutations of neural activity. However, it is the brain state that possesses causal efficacy, not the mental states of perception, reflection, and deciding. However, viewed holistically, it is the *mental-material union* that exemplifies synergy.[57]

Through this vast variety of combinations and permutations of neural activity with concomitant mental states, the nexus in the causal time-line becomes highly concentrated in the neural network of the brain. As such, the brain could be described as *initiating* this *deciding* mental state, albeit with prior ignition from an earlier part of the causal nexus; such as, olfactory nerve stimulation from chemicals in the air caused by something else, and so on in the nexus. Moreover, although much of the neural network structure is pre-determined by genetics, it is also determined by experience, the plasticity of the brain is undergoing constant change. Such a malleable neural network helps define the *identity of a person*, particularly reasoning and no more so than moral reasoning. In this sense and given the initial causal trigger, it is the *person* who initiates the *deciding* mental state; immediate agency predominantly resides with the person. Thus, the above revisionist (mental[*will*]-material[*free*] union) conception of free-will is commensurable with the requirement that it is the *power of the person to initiate decisions* that delineates free-will, albeit still within the causal nexus.

Harry Frankfurt (1971) takes the free-will explanation further than just initiating decisions, differentiating between spontaneous desires and volitions to act upon those desires after reflection. The former he terms first-order wants, the latter second- order wants.[58] Frankfurt argues that it is the second order wants that define free-will.[59] Frankfurt's analysis is compatible with the nexus-concentration argument in the brain. Spontaneous desires are mental state properties supervening on relatively simple brain state events that perhaps have an extrinsic cause. For example, the perceptual brain-state and emergent mental state of smelling baking bread causing the spontaneous brain state and emergent mental state of desiring to eat the bread; a first-order want. The mental states of reflecting, deciding, and acting[60] involves considerably more brain event activity than the spontaneous state of affairs; a second-order want where the brain and hence the person can be described as predominantly initiating subsequent action despite the initial perceptual cause.

Thus, although a folk perspective would have us think of free-will as a mental property descriptor, it is a property that cannot be isolated from its relevant material property. Together, they can be justifiably described as free and cause effects of their own, both within the brain and, through the causal nexus, to the wider world outside of the brain.[61]

Plantinga's Modal Defence

Given the attribute of omni-benevolence, Alvin Plantinga (1982) states that "A world containing creatures who are sometimes significantly free (and freely perform more good than evil actions) is more valuable, all else being equal, than a world containing no free creatures at all." [Plantinga, 1982, 85] Based upon this premise, Plantinga provides a free-will defence based upon an omni-being either *strongly actualizing* a possible world or, alternatively *weakly actualizing* a possible world. If an omni-being strongly actualizes a possible world which contains a contingent state of affairs where John *freely* chooses to do a malevolent act, then in effect, the omni-being *determines* John's action and the act ceases to become a free choice that is completely determined by John. Alternatively, if the omni-being strongly actualizes the maximum state of affairs in that possible world necessary for the actualization of that world, but excluding the contingent outcome of John's choice then, in effect, two possibilities emerge; John will choose to act malevolently or will choose not to act malevolently. Plantinga terms this *weak actualization*. Thus, in such a world that is weakly actualized by the omni-being, it is John and not the omni-being that, *ceteris paribus*, determines the moral choice.[62]

Given the attribute of omni-benevolence, it is possible that it is not within the omni-being's power to actualize a world with moral good and not moral evil.[63] The same principle applies for any quantity of moral evil; thus, it is not "…within the power of [the omni-being] to create a world as good as the actual world, but less evil." [Plantinga, 1988, 78][64]

There is also a free-will theodicy and the free-will debate is one area where defence and theodicy overlap. Nevertheless, a theodicy engages with the plausibility of free-will as a justification for evil and God to coexist, whereas Plantinga's defence aims to demonstrate that God and evil are not logically incompatible as John Mackie (1955) forcefully argues.

Counter Argument

Plantinga's argument is about evil per se not about specific cases of evil; horrendous evils like 'Bambi' and 'Sue' that may defeat, in terms of severity, the high value ascribed to free-will and hence be gratuitous. Thus, God could strongly actualize a state of affairs that includes the contingent act of John freely choosing not to engage in gratuitous evil, but does not include John freely choosing to engage or not to engage in non-gratuitous evil. This preserves the free-will of John in respect to evil per se, preserves the greater good of free-will over such evil, but eliminates gratuitous evil.[65]

Robert Mesle (1991) questions the high value given to the free-will defence of moral evil. He argues that free-will is not valued because of malevolent choices, but because of the opportunity for benevolent choices. It is the choice between goods that is valued not between depraved choices.[66] However, even though free-will can result in malevolent choice, without free-will, moral choices would not be available to us.[67]

§ SECTION THREE: THEODICY

Theodicies[68] differ from the defences above in so far as they provide specific reasons why God and evil coexist, whereas defences limit themselves only to the possibility of there being a reason. There are overlaps between defences and theodicies, the free-will defence/theodicy being a case in point.

Given the assumption in theodicies that God exists, they have no probabilistic confirming or disconfirming application in the likelihood function $Pr(e|\neg p \; \& \; k)$. However, justifiable theodicies can provide probabilistic confirmation in the likelihood function $Pr(e|p \; \& \; k)$ in so far as theodicies are intended to provide an explanation for God permitting the existence of evil in the world.

Question Begging

Theodicies are justificatory in form; indeed, *theodicy* is derived from the Greek *theos* meaning God and *dikē* meaning justice. Thus, theodicies are viewed as justifying the coexistence of God and evil for the greater good of, for example, free-will or soul-making. If such justification is moral justification, theodicies are question begging as God's existence is pre-assumed and the omni-predicates ensure God's moral perfection; thus, God's moral perfection is justified by the theodicy that has already assumed God's moral perfection. However, justification can also imply good reasons, and to avoid the charge of circularity, justification should be employed in this form.

Circularity still lurks with religious theodicies as the justification cited can entail God's existence. For example, the Fall in Genesis explains why God expelled Adam and Eve from the sanctity of Eden to justifiably suffer, along with their descendants, the natural and moral evils of the world outside Eden. Thus, the *existence of God* sits within the very reason propounded for evil and God to coexist leading to the conclusion that *God exists*.

Theodicies span a range of benefits acquired in the face of suffering; the development of valued personal characteristics such as virtue and fortitude, the value of free-will, and the value of stable natural laws. There are also religious theodicies, the most common one in the Christian world is the fall of man in Genesis and the forgiveness of sins through the sacrificial death of Jesus. I do not intend to address religious theodicies because of their inherent circularity.

Soul-Making Theodicy

This theodicy is associated with John Hick (1974), whereby persons undergo spiritual development through free-will actions, particularly overcoming temptations, enabling eventual communion with God. Such opportunity for continual spiritual development, termed soul-making, is viewed as a greater good, justifying God permitting moral and natural evils in the world.[69]

Counter Argument

It could be argued that some suffering is so horrendous it exceeds the level required for spiritual development, particularly if it ends in the death of the victim as no development can occur if death swiftly follows the suffering. Also, unless it is conceded that animals also have souls, the apparent gratuitous suffering of 'Bambi' cannot be justified on this theodicy. Likewise, the suffering and death of children such as 'Sue,' where the opportunity for life-long spiritual development has been removed is difficult to reconcile with this theodicy.

This theodicy also boarders on the religious because the objective of soul-making is communion with God; it is, therefore, question begging.

Free-Will Theodicy

Mostly it is libertarian free-will that is employed in the free-will theodicy and it is viewed by the supporters of the theodicy as of such great value that it is better to suffer the evils that result from it than to lose it. Supporters of this theodicy also assume that God also values libertarian free-will and hence being free even from His intervention; hence, the quantity and intensity of moral evil in the world.

Counter Argument

Libertarian free-will is not as widely accepted as that of compatibilism and compatibilism is itself compatible with the free-will theodicy as God does not have a necessary causal relationship with persons such that their free-will decisions are not actually free; God can choose not to intervene because of the high value of free-will. This places a burden of responsibility on God, but it does so even with the libertarian version. For God could intervene when evil becomes too horrendous (see the counter to natural law theodicy below). Further, persons and animals could have a bio-limiter in their constitution so that the level of suffering they can inflict is limited. In addition, a free-will theodicy does not encompass natural evil.

Free-Will and Horrendous Evil

Richard Swinburne (1998) suggests that the possibility of engaging in severe moral evils are important to ensure meaningful responsibility for the welfare of other persons through a wide range of malevolent and ameliorative actions. Limited scope in the possible expression of moral evil reduces the scope and value of the responsibility vested in the gift of free-will.[70]

Counter Argument

This argument does not apply to the problem of horrendous *natural* evil, although Swinburne does address this (see virtue theodicy below). Also, wide scope in moral value does not entail wide scope in malevolent action. Wide scope in possible benevolent action, that could be withheld or provided, would also ensure a wide scope in moral value without the need for inflicting horrendous moral evil.[71] Moreover, Swinburne's argument relates to the mental state of the possible evil doer rather than the act itself. Consequently, the desire to inflict severe harm is all that is needed and the greatest personal benefit is overcoming the desire. Thus, a prick of conscience caused by God could stop the act, but the benefit of extreme malevolent desire overcome is gained.

Virtue Theodicy and Natural Evil

Richard Swinburne (2004), suggests the possibility of greater good resulting from natural evil. He introduces a virtue theodicy whereby natural evils provide the opportunity to express the goods of compassion, sympathy, and courage to overcome them. They will, of course be accompanied by temptations, but these require even greater fortitude to overcome them.[72]

Counter Argument

An alternative to actual natural evil would be the divinely created illusion of natural evil saving the reality of suffering.

Reply

However, this would be an illusion that could not be fulfilled as at some point in the illusionary nexus of cause and effect the evil illusion would eventually become apparent and all natural evil thereafter viewed as a tongue-in-cheek deception. Swinburne argues that actual experience or observational experience of natural evil is necessary to ensure predictability in nature and strategy to avoid future repetition.[73]

John Mackie (1955) questions whether evil can ever produce greater goods. He introduces the notion of contrasting levels of good and evil. The first-order evils of pain and suffering contrast with the first-order goods of pleasure and happiness. Reaction to first-order evils can produce second-order goods that outweigh the first-order evils; for example, benevolence, bravery, and sympathy. However, given free-will, second-order evils may result, instead of the second-order goods; evils such as malevolence instead of benevolence, cowardice instead of bravery, and callousness instead of sympathy. Such second-order evils can give rise to additional first-order evils of pain and suffering. Thus, the quantity of evil increases; original and additional first-order evils plus second-order evils. These second-order evils trigger the possibility of third-order goods as well as third-order evils. If so the stock of evil increases even more. Mackie argues that we are faced with the possibility of an infinite regress and an ever-increasing surplus of evil.[74]

Counter Argument

Are the "joys and sufferings of life…constructed like a multi-story car park [?] In real life, virtues or 'second order goods' respond both to pains and to vices and there is no 'further problem to be solved' with a different level of goods." [Griffith-Dickson, 2000, 171]

David O'Connor (1998) believes that Swinburne is being excessive in his virtue theodicy. Natural evil, he argues, falls into two categories: evil resulting from natural processes, such as volcanoes and earthquakes and evil resulting from non-natural processes, such as accidents and negligence. He believes that the non-natural category would be enough to satisfy the greater good of virtue. Consequently, the natural processes form of evil is excessive and inconsistent with the existence of God; O'Connor terms this counter argument *the reformed logical problem of natural evil*.[75]

Reply

I am not convinced by O'Connor's argument that evil resulting from non-natural processes such as accidents are frequently not independent of natural processes. The suffering caused by falling off a bicycle due to excessive alcohol involves gravity, the effects of alcohol on the brain and the fragility of the body.

The Theodicy of Divine Withdrawal

One theodicy that answers the above free-will defeaters and the problem of natural evil is that of divine withdrawal. God's essential being is expressed by the omni-predicates making anything but perfection in His presence unacceptable. This includes both the natural world and persons. Assuming free-will is important to God, and given the necessity of perfection in God's presence, such presence would only permit them to do what is right, but free-will entails them being able freely to do both right and wrong. Thus, the instantiation of free persons must be distanced from God. Once the creation of the universe is set in motion, God must withdraw to permit persons to emerge freely from the physics of the creation; an evolution of free-will. Thus, God cannot intervene by eliminating horrendous evils, as His presence during the act would ensure all evil is eliminated and free-will would be lost. God cannot intervene in natural disasters because His presence in the act would demand perfection in everything and free-will would again be lost. The idea of bio-limiters presumes specific laws of physics that ensure the evolutionary emergence of the limiters and that seems implausible. If such limiters were to be divinely engineered, then God is choosing to restrict free-will action. The argument is based upon the idea of God making things that make themselves, in this case starting off the universe with the natural laws as nomic-seeds for its future and then withdrawing. This is somewhat like the theodicy of David Birnbaum (1989) who refers to *zimzum* (or *Tsimtsum*) or divine contraction from the here and now[76] leaving man to engage with evil directly to fulfill his quest for potential.[77] In Birnbaum's view divine contraction can be partial in the sense of proportional contraction,[78] but divine perfection is not itself partial and imperfection would be ameliorated even with a partial theophany.

Counter Argument

The most obvious problem with this theodicy is that God is an absent God by necessity. This raises the problem of how He engages with His creation. In the next chapter, I argue that there is probabilistic confirmation that God does engage with persons, evidenced by religious experience.

Strictly speaking, this is not a theodicy in so far as God and evil are not co-existing as God has withdrawn His presence.

Reply

One possible explanation is that the connection is not God directly, but some sort of intermediary (possibly the Holy Spirit) or that connectivity is enabled in some way without divine presence. An analogy would be the effect of the sun; we feel its warmth, see its light and our planet is held by its gravitational force, yet it is distant from us and its close presence would destroy us. Some form of connectivity may permit this state of affairs to be termed a theodicy; notwithstanding, we should not be strictly bound by definition.

Natural Law Theodicy

This theodicy is based upon the importance of regularity in nature. Irregularity would stymie knowledge, understanding, skill, prediction, planning, and effective action; neither man nor beast would prosper or even evolve. The finely tuned laws of nature ensure such regularity but bring with it the suffering of natural evils such as tsunamis, earthquakes, and fires to name just three of a vast number of natural disasters. The theodicy claims that God is justified in permitting natural evil for the sake of the natural laws and the regularity in nature that follows. It would be a greater evil if it were otherwise.

Counter Argument

Occasional divine intervention in the worst disasters would not eliminate the stability that the natural laws bring. In addition, the natural laws would function in a world with different boundary conditions such as one with no bacteria or viruses that cause very serious diseases.

Reply

This would be no different from preventative policing, other than it would, of course, reveal God's presence or at least the existence of a supernatural force that defies natural laws or at least prevents belief in predictable regularities. The result is that horrendous evils would enter the realm of science fiction; imaginatively conceivable but practically impossible. Also, the natural law theodicy does not account for moral evil.

Alston's Theodicies

William Alston (1991) approaches the problem of gratuitous evil through the application of multiple theodicies. Alston adds to the 'Bambi' and 'Sue' examples with 'Sam,' an adult sufferer from a painful and lengthy illness. The theodicies selected apply as follows: 'Bambi' the natural law theodicy, 'Sue' the free-will theodicy and the martyrdom theodicy[79] and 'Sam' the soul-making theodicy. Each event is, *prima facie*, inscrutable because of the human ignorance of the justification for God to permit such events; the theodicies only representing conceivable justifications. The strength in Alston's approach is that he does not rely on one theodicy, but a conjunction of many that collectively provide a challenge to the atheistic sceptic.

Alston sidesteps the charge of begging the question by not trying to establish a theistic conclusion, but rather defending against arguments that attempt to establish an atheistic conclusion.[80] This is a further example of the overlap between defence and theodicy.

Counter Argument

A conjunction of theodicies increases the probability of failure of the argument as only one must be false for the conjunction to be false and there are counter arguments above for each of the theodicies quoted other than martyrdom, which I have not addressed. Theodicies do not easily lend themselves to falsification as they are broad and invoke the will of God; this is a weakness in the theodicy argument generally.

Confirming or Disconfirming the Evidence on the Negation of the Proposition

This is a clear and simple argument for $\Pr(e|\neg p \ \& \ k)$. God is assumed not to exist and background knowledge k comprises naturalistic and non-theistic explanations for the existence of moral and natural evil in the world. 'Red in tooth and claw' exemplifies the context of the animal world where predators hunt prey and natural evil, in terms of animal suffering, is common. Geological forces cause volcanic eruptions, earthquakes, and tsunamis, where both humans and animals fall victim to these natural forces. The process of evolution has produced bacteria and viruses that have caused plagues that have killed millions. People over eons have engaged in extreme violence against one another in wars and criminal aggression for a whole host of reasons. Indeed, the world has been and is teeming with natural and moral evil that can be explained by non-supernatural events. Moreover, even the possibility of divine control has been assumed non-existent in the likelihood function $\Pr(e|\neg p \ \& \ k)$; thus, e is what we would expect given the prevailing naturalistic context.

EVALUATORS' CONCLUSION

The problem of evil is undoubtedly the most problematic evidential argument for the theistic proposition *God exists*. The deductive argument for the non-existence of God in the face of evil does not hold. However, the inductive and probabilistic arguments have some traction, but falter using logical rather than subjective probability together with some negligence concerning the principle of total evidence. The theistic apologists generally resort to a defence by claiming ignorance of greater goods known unto God. Notwithstanding, theodicies do have probabilistic confirming force with the theodicy of divine withdrawal, combined with the free-will defence, leading the pack. By contrast, naturalist explanations for evil are very persuasive and suggest a high value for the likelihood function $\Pr(e|\neg p \,\&\, k)$.

Evaluators' Judgement

1. *Atheist's view*: a low value for $\Pr(e|p \,\&\, k)$, with scepticism concerning theodicies. Such scepticism will create doubt in the justification of theodicies as auxiliary assumptions in *k*. (See Appendix 1) In the context of subjective probability, if such doubt about an auxiliary assumption exists, then its confirming force on the probability of the evidence will, in the eyes of the sceptic, reduce. A high value for $\Pr(e|\neg p \,\&\, k)$ given auxiliary assumptions concerning the dark side of human and animal behavior, as well as the disordered nature of the material world.
2. *Agnostic's view*: a low value for $\Pr(e|p \,\&\, k)$, with some scepticism concerning theodicies and a high value for $\Pr(e|\neg p \,\&\, k)$.
3. *Theist's view*: a medium to low value for $\Pr(e|p \,\&\, k)$ with a positive view of theodicies and a high value for $\Pr(e|\neg p \,\&\, k)$.

Consensus

A consensus would be expected with an acknowledgement of the possibility of unknown greater goods, but known unto God and the theist's attraction to theodicies, particularly the divine withdrawal theodicy. Some compromise through discussion leading to consensus as follows:

$\Pr(e|p \,\&\, k) = 0.3$
$\Pr(e|\neg p \,\&\, k) = 0.8$
$\Pr(e|p \,\&\, k)/\Pr(e|\neg p \,\&\, k) = \lambda = 0.38$

Prior Probability

From the previous chapter, the following priors apply to this evidential element: atheist: 0.17; agnostic: 0.64; theist 0.91.

Bayesian Conditionalization

The existence of moral and natural evil in the world is evident. Also, the posterior values are not so extreme as to cause a negative reaction from the syndicate, particularly as they know there are other evidential elements to be considered. Therefore, it is reasonable to assume conditionalization by the hypothetical syndicate members. The posterior probability values below then become the priors for the next evidential element.

Posterior Probability

From Bayes' theorem (likelihood ratio form):

$$\Pr(p|e \;\&\; k) = \frac{\lambda \times \Pr(p|k)}{[\lambda \times \Pr(p|k)] + [1 - \Pr(p|k)]}$$

Substituting the three prior probabilities of the hypothetical syndicate members into the theorem:

Atheist

$$\Pr(p|e \;\&\; k) = \frac{0.38 \times 0.17}{[0.38 \times 0.17] + [1 - 0.17]} = 0.07$$

Agnostic

$$\Pr(p|e \;\&\; k) = \frac{0.38 \times 0.64}{[0.38 \times 0.64] + [1 - 0.64]} = 0.40$$

Theist

$$\Pr(p|e \;\&\; k) = \frac{0.38 \times 0.91}{[0.38 \times 0.91] + [1 - 0.91]} = 0.79$$

NOTES

1. I have entitled the chapter *The Bookie's Favorite* because of the allusion to the betting Coherence argument of Frank Ramsey (1926) and Bruno De Finette (1930) [*Subjective Probability*; Chapter 2]. In this case the bookie represents the atheist camp and the punter the theist camp.

2. Griffith-Dickson G. (2005); p. 362.

3. Epicurus was a Greek philosopher (341–270 BC) and founder of Epicureanism, a philosophy based on the premise that that pleasure is the primary purpose of life and that one's beliefs should be founded on observation and logical deduction.

4. 'Bambi' was first introduced by William L. Rowe (1996)^{A}; p. 4 and 'Sue' by Bruce Russell (1989)—*The Persistent Problem of Evil*—in Faith and Philosophy 6; pp. 121–139.

5. The deontological argument in ethics claims that actions are right or wrong in themselves irrespective of their consequences: Davis N. (2003); pp. 205–218.

6. Tooley M. (2015); Sec.3.2.1.

7. Hume D.[1748] (1993); IV; 2 and Howson C. (2011); p. 125.

8. Draper calls this proposition "the hypothesis of indifference" (not to be confused with the principle of indifference of probability theory).

9. Draper P. (1996); pp. 12–29.

10. Note the possible inductive relationship between P and Q introduced in the deontological argument above (*The Inductive Argument*).

11. Rowe W. L. (1996)^{B}; p. 265f.

12. Otte R. (2002); p. 149.

13. Gillies D. (2003); pp. 37–49.

14. Tooley M. (2015); Sec.3.4.2.

15. Rowe W. L. (1996)^{B}; p. 265.

16. Richard Otte (2002) also argues that, as well as P, there is direct evidence unaccounted for in Rowe's analysis argument; i.e., no God-justifying goods for the evil even if God does not exist. Otte provides an alternative G-disconfirming analysis to Rowe considering this additional evidence.

17. Rowe acknowledges this proposition termed P*, but states it should not be confused with P; Rowe W. L. (1996)^{B}; p. 283.

18. *Ibid*; p. 270.

19. Plantinga uses P* instead of P^{A}; I have used the latter so as not to confuse the proposition with my earlier use of P*.

20. This is more of a metaphysical entailment; that is, the divine attributes ensure the truth of P^{A}.

21. Plantinga A. (1998); p. 536f.

22. Proof:.

    1. *Entailment*: $\neg G$ entails P.
    2. *Assumption*: $Pr(P) < 1$.
    3. *Conditional probability*: $Pr(\neg G|P) = Pr(\neg G \& P)/Pr(P)$.
    4. *From axiom 2*: $Pr(\neg G \& P)/Pr(P) > Pr(\neg G \& P)$ (ensures $Pr(.) \leq 1$).

5. *From 1*: Pr(¬G & P) = Pr(¬G).
6. *From 4 & 5*: Pr (¬G & P)Pr(P) > Pr(¬G).
7. *From 3 & 6*: Pr(¬G|P) > Pr(¬G): confirmation (*Ibid*; 537f).

23. *Ibid*; 240f. For Rowe's reply to Plantinga: Rowe W. L. (1998); *Reply to Plantinga*—in NOÛS 32:4; pp. 545–552.

24. This does assume that omnipotence and omniscience entail omni-benevolence; Ch. 1; Omni-benevolence.

25. Rowe W. L. (1996)[A]; p. 6ff.

26. Noseeums are midges; tiny biting flies that are so small you 'no-see-um.'

27. Wykstra S. (1990); p. 155f.

28. Rowe W. L. (1991); pp. 77–79.

29. Wykstra S. (1990); p. 132f.

30. Howard-Snyder D., Bergmann M. & Rowe W. L. (2001); p. 148f.

31. The Argument from Providence; Section 3, Chapter 3.

32. Hasker W. (1992); pp. 23–44.

33. O'Connor D. (1998); pp. 57–76.

34. *Ibid*; pp. 56–63.

35. Exploring the issue of moral responsibility and free-will is beyond the scope of this book, but I would direct the interested reader to Caruso G. D. (2013) (Ed.).

36. Such a conception is known as epiphenomenalism: Kim J. (1984); p. 257f.

37. Harris S. (2008)—*Free-will*—Simon & Schuster.

38. Determinism is the thesis that all events in the world are caused by past events within a causal nexus and are governed by nomological laws such that all future events are determined even apparent free decisions. This is also known as strong or hard determinism; soft determinism allows for free-will to operate within the constraints of determinism. For a more detailed analysis see Van Inwagen P. (1975); pp. 46–49.

39. Cartesian dualism is the paradigm of substance dualism: Maslin K. T. (2007); pp. 38–41. For a defence of substance dualism see Foster J. (1996).

40. Bradley R. D. (2015); Ch. 11.

41. This is an across-worlds supervenience notion, i.e., *it is not the case* that there are two possible worlds where the material properties of a brain event are the same in each, but its mental properties differ. Davies M. (1995); p. 270f.

42. Some philosophers give mind more independence than that of being a property of the brain, with the mind-brain causal interface being an inexplicable brute fact: Paterson R. W. K. (1995); p. 98f and Plantinga A. (2007).

43. Davidson D. (1970); pp. 247–256.

44. For purposes of clarity I refer to a *brain state* as that state the material brain is in at a moment and exhibiting *mental properties*; e.g., feeling pain, and *material properties*, e.g., c-fibers firing.

45. Greenfield S. (2000); p. 2. No doubt there is, or will be, a neurological explanation for this brain-property/mental-property union; an issue beyond the scope of this book.

46. It is not exactly synchronous; the brain state is slightly in advance of the mental state: Soon C. S. et al. (2008) and Libet B., Gleason C., Wright E. and Pearl D. (1983).

47. Kim J. (2007).

48. This lack of causal efficacy raises the issue of the purpose of a mental state, which likely has evolutionary value to exist at all. Although beyond the scope of this book, a good starting place for the interested reader would be Dennett D. C. (1993); pp. 171–226.

49. This, of course, does not account for the material spontaneity expressed in quantum mechanics.

50. Incompatibilism is the position taken by philosophers who argue that the notion of free-will is incompatible with causal determinism; this contrasts with Compatibilism. For an incompatibilist argument see Van Inwagen P. (1975).

51. Vargus M. (2013); p. 182f.

52. Chisholm R. M. (1964); pp. 24–35.

53. Richard Swinburne takes a different position, arguing for a libertarian rather than compatibilist explanation to exonerate God from being responsible for the malevolent acts of persons: Swinburne R. (1998); p. 34.

54. Ayer A. J. (1954); p. 15.

55. c-fibers and d-fibers are imaginative representations for illustrative purposes only.

56. *Causa sui* means an uncaused cause or first cause, i.e., the brain leaps into spontaneous action without a prior cause. This would challenge the determinist's case against free-will, but it is not a popular position at present: Strawson G. (2008); pp. 319–335.

57. Llinás and Paré (1996) have hypothesized such a synergetic causation of consciousness through the synchronization of neuron mass from different brain regions.

58. According to Frankfurt, beings restricted to first order wants are either animals or human wantons; Frankfurt H. (1971); p. 86f.

59. *Ibid*; pp. 89–92.

60. Frankfurt argues that the action element of the sequence may be restricted, yet freedom of will remains. *Ibid*; p. 90 This also corresponds with the nexus-concentration argument in the brain.

61. Koch C. (2012); Ch. 7 for neuroscientist's compatibilist argument.

62. This is a libertarian free-will argument, i.e., John's will is undetermined by extrinsic factors. Above I argue for a compatibilist position, i.e., that John's will can still be free in a deterministic context; hence, my use of *ceteris paribus*.

63. Plantinga A. (1988); pp. 76–78.

64. For a more rigorous and detailed account of Plantinga's modal problem of evil argument see Plantinga A. (1982); Ch. IX and for a simplified version see Plantinga A. (1988); Part 1.

65. Tooley M. (2015); Sec. 1.3 also Rowe W. L. (1991); p. 75.

66. Mesle C. R. 1991 p. 42f.

67. Griffith-Dickson G. (2000); p. 233.

68. The term 'theodicy' was first coined by Gottfried Leibniz (1710) (1951).

69. Hick J. (1981); pp. 265–281.

70. Swinburne R. (1998); p. 147ff.

71. This is a deontological view in so far as providing or withholding assistance is right and wrong in itself, whereas a consequentialist view is that what is right or wrong is the outcome. The view chosen changes the force of this defeater.

72. Swinburne R. (2004); p. 240f.
73. *Ibid*; p. 248.
74. Mackie J. (1955); pp. 30–32.
75. O'Connor D. (1998); pp. 77–91.
76. Birnbaum presents the idea of perfection being only within God; creation, by definition, being outside of Him is therefore imperfect; Birnbaum D. (1989); p. 28.
77. *Ibid*; 80.20.
78. *Ibid*, 900.20c & 900.11c.
79. Alston views the martyrdom theodicy as a "benefit to human beings other than the sufferer or to humanity generally", i.e., for the greater good of aspiration in the onlooker or repentance in the perpetrator: Alston W. (1991); p. 114.
80. *Ibid*; p. 104.

Chapter 5

A Short Head

Evidence from the Divine/Human Relationship

Progressing through the evidential elements that confirm and disconfirm the truth of the theistic proposition *God exists*, brings us to the evidence from the divine/human relationship, with the first evidential element for consideration being religious experience. Again, the title of the chapter alludes to horseracing to reflect the betting platform of subjective probability theory.[1]

Given the hiddenness of God or from the atheist's perspective, the non-existence of God, His relationship with us or his illusionary relationship with us is of paramount importance. If there is evidence of a relationship, then this is confirmatory in a probabilistic way. If such evidence is self-delusional, then the evidence is not confirmatory.

Although there is a standard set of evidential elements in this chapter, the reader may think of more and, if they have participated in the subjective Bayesian process up to this point, they can continue with their own additions at the end of this chapter. I intend to employ the three probability values specified at the end of Chapter 4 as the priors for the theistic proposition in this chapter.

The Evidence

There are three sections comprising this chapter, each addressing an evidential element; religious experience, miracles, and the *consensus gentium* argument

As in the previous two chapters, the Bayesian methodology I employ is through the application of the likelihood ratio $\Pr(e|p \,\&\, k)/\Pr(e|\neg p \,\&\, k)$.

§ SECTION ONE: RELIGIOUS EXPERIENCE

This is the sixth evidential element in the set and is problematic from the outset, as the term *religious experience* can be misleading as it does not refer to experience of a religion; for example, listening to or reciting the Nicene Creed. Such a recital can be edifying and apologists may state that such edification is an example of religious experience.

Religious experience is not the experience of a religious activity per se, but the experience resulting from *inter alia*, that activity. A good analogy is the edifying pleasure felt listening to music or viewing art or supporting a favored sports team and other activities. The supposed cause, in the case of religious experience, is not an aria, impressionist painting, or skillful striker but God.

To describe an unusual episode as *religious experience* and then proceed to use this as evidence for the existence of God begs the question, as use of the term *religious* can presuppose God's existence. Thus, care must be taken in its application and what follows will highlight the bias that can shadow its misuse.

The Problem of Interpretation

Prima facie, the occurrence of religious experience seems an obvious evidential element that probabilistically confirms the theistic proposition *p*. However, such a self-evident conclusion is deceptive, primarily because of the circularity, but also there is a related *problem of interpretation.*

It could be argued that predicating an unusual experience with *religious* is simply interpreting the experience in a particular way; that is, in a theistic way, whereas an atheist would probably interpret the same experience in a naturalistic way; for example, an illusion. Thus, the experience could be very similar, but the interpretation distinct.[2] The problem facing the apologist is to discover which interpretation is right. If no clear explanatory advantage can be found in either interpretation, then an impasse arises and the hope of a complete consensus in an intersubjective probabilistic analysis is doomed.

An Uncanny Experience

The following analysis engages with this potential impasse, and to provide impartiality in the analysis, I will predicate an inscrutable, unusual, or mysterious experience as an *uncanny experience*. If interpreted as being divinely caused, then the interpretation is *religious*, if not it is naturalistic. I could use *religious experience* with a caveat to the above effect, but it may surreptitiously introduce bias; therefore, I shall keep its use to the minimum. All the authors referred to tend to use the expression *religious experience*; other than in direct quotations, I change this to *uncanny experiences presumed to be divinely caused* or something similar. This approach does appear to be incongruous with current literature on religious experience. However, when making probabilistic judgements of the subjective kind, loaded descriptors such as *religious* can prejudice that judgement, despite having little or no effect in other areas of the philosophy of religion.

CONFIRMING THE EVIDENCE ON THE PROPOSITION

I will now consider arguments that attempt to tilt the balance in favor of the theistic viewpoint; that is $\Pr(e|p \,\&\, k) > \Pr(e|\neg p \,\&\, k)$ where e is evidence of an uncanny experience. In the same style as the previous two chapters, I will provide some counter arguments and replies for consideration.

Divine Hiddenness

Given the proposed hiddenness of God because of His intentional withdrawal from our universe as suggested in the previous chapter,[3] it could be argued that it is more probable than not that there will be cases of divinely caused uncanny experiences, for if God is not manifestly present in our world, then without some form of connectivity God would be distant and impersonal. An analogy would be a loved one on a long business trip; it would be surprising if no text, e-mail, or letter were sent or a telephone call made. The lack of such connectivity would be indicative of love lost. Likewise, no human and divine connectivity would also be indicative of love lost; that is, contrary to the divine attribute of omni-benevolence.

Counter Argument

The obvious counter is that the argument presupposes the existence of God. To then go on to use divinely caused experience as evidence in support of God's existence takes the previously mentioned circularity to new depths. Moreover, it could be argued that God is beyond contact and therefore, to believe that such uncanny experiences are divinely caused, is delusional. The argument also presupposes the veracity of the divine withdrawal theodicy.

Doxastic Practice

Given that I am, unusually, considering uncanny experience as evidence that probabilistically confirms the truth of the theistic proposition *God exists*, the issue arises of whether it can be considered as evidence at all. William Alston (1993) argues that treating such experience as evidence is not so very different to other forms of experiential-based knowledge; that is, beliefs formed from other well established practices that are considered rational.[4] Alston terms this *doxastic practice*;[5] that is; belief forming systems comprising inputs (sense experiences, for example) and outputs (beliefs).[6] A paradigm being beliefs established from scientific practice. Such beliefs attain warrant by reliance on other practices; in particular, observation, deduction, and induction. However, scientific practices also attain warrant from assuming the truth of premises derived from scientific practice itself; that is, assumptions that beg the question. Despite this circularity, such assumptions are justified through established practices over time and by *intrinsic consistency* within the scientific network of beliefs and *extrinsic consistency* with other established belief networks.[7] Perhaps this provides some respite to the problem of circularity inherent in the title *religious experience*.

Counter Argument

The obvious problem with this argument is that there exist two interpretations; that is, a theistic one and naturalistic one. The real issue is what interpretation is correct; perhaps they both are acceptable for similar experiences; what Alston calls an 'overrider.'[8] Further, there exists a wide variation in interpretation of uncanny experiences even between members of the divinely caused interpretive set. This is like disagreement in eyewitness reports of a robbery, for example. It could be argued that, like the loss of credibility with such eyewitness reports, the wide variety of assumed divinely caused experiences loses credibility; what Alston calls an 'underminer.'

Reply

Many uncanny experiences are personal and, if divinely caused, individual variation in the contents of reports are to be expected as in any person-to-person connectivity.[9] Moreover, interpretation is not culturally neutral and consequently, culturally influenced interpretations are also to be expected.[10] Although variations do cluster into types as classified by Alister Hardy (1979)[11] and Caroline Franks Davis (1999), to argue that all such experiences have a divine cause is questionable, particularly given alternative naturalistic explanations.

Foundationalism

Alvin Plantinga (1981) provides an argument from foundationalism[12] whereby he accepts that while sense experience per se is not infallible, we build a cognitive superstructure of the external world upon it. Beliefs formed from sense experiences are treated as properly basic;[13] that is, not requiring justification by argument despite such fallibility. In the same vein, Plantinga argues that God manifests Himself to us through uncanny experience and properly basic religious beliefs are derived from such sense experiences; for example, 'God is speaking to me.' Plantinga argues that such a religious belief assertion is properly basic because it sits, *ceteris paribus*, within the set of what counts or what is putative as a properly basic belief, unlike, for example, "…the Great Pumpkin returns every Halloween?" [Plantinga, 1981, 101] Thus, argues Plantinga, based upon paradigm examples, the experient can justifiably hold a divine experiential belief on the grounds of it being properly basic; that is, taken at face value and not based upon other beliefs.[14]

Counter Argument

Again, there is a prior assumption of or at best an allusion to God's existence and that He is the cause of the uncanny experience. The uncanny experience may be basic, predicating it with the descriptor *religious* is bias. On this basis, it is inappropriate to employ it as evidence to support the theistic proposition. Plantinga's intention is to argue that holding a religious belief based upon uncanny experience interpreted as divinely caused is not necessarily a case of noetic impropriety. Nevertheless, an uncanny experience can elicit differing interpretations from a theist on the one hand, and an atheist on the other. The properly basic belief formed from such an experience could be, in the former case, *God is speaking to me* and, in the latter case, *I imagined a voice in my mind's ear*. The experiences can be properly basic, but the interpretation varied and an epistemic impasse concerning the cause of the belief results.

Credulity, Testimony, and Busy Bees

Richard Swinburne (2004) provides an argument justifying the interpretation of an uncanny experience as being divinely caused through the *principle of credulity*. Thus, if a person perceives some uncanny experience to be a divinely caused, then it probably is.[15] The underlying notion of the legal witness system is also based upon this principle. Thus, *prima facie*, the perception is justified on this principle, but it can be undermined by contrary evidence or argument. Swinburne supplements the principle of credulity with the *principle of testimony*; that is, people normally tell the truth.[16] Thus, combining the two principles, if someone reports having a divinely caused uncanny experience, then *ceteris paribus*, they probably did.

Jerome Gellman (1997) provides a similar argument to Swinburne's principle of credulity arguing that the best explanation of a person's reported experience of an object or kind of object is, *ceteris paribus*, 'that person experienced the object' as opposed to anything else. Gellman terms this principle the Best Explanation of Experience (BEE)[17] and that the more people who have the same experience the more veridical the report: Strength in Number Greatness (STING).[18] Clearly, STING probabilistically increases the veracity of the experience; for example, if six people saw Jones robbing the bank, then *ceteris paribus*, it is more justified to believe that Jones did rob the bank than if one person saw Jones.

Counter Argument

Notwithstanding the buzz-words, Swinburne's principles and Gellman's arguments are convincing when applied to eyewitness accounts of material events; for example, a robbery, but the application to uncanny experience is more problematic. This is because the experience may justifiably be thought of as divinely caused by one person and justifiably naturalistically caused by another; there is a strong interpretive element involved. The principles and arguments apply equally as well to both these conflicting interpretations. The important question is why one person predicates an experience as *religious* and another as *natural*? If a religious interpretation is a consequence of a particular world-view, then possibly the interpretation is bias, but then the same applies to a naturalistic interpretation; an impasse. Further, STING weakens when applied to uncanny experiences thought to be divinely caused as most specific reports are personal.

Introspection

Keith Yandell (1994) tackles the evidential nature of uncanny experience thought to be religious by assuming, for the sake of argument, that the experience is an experience of a supernatural being. Yandell suggests that the experience counts as evidence for the existence of God provided the experient is not deluded; that is, she does not believe she would have an experience of a supernatural being whether there is one or not and that, if her conclusions based upon her experience or type of experience are false, she does not believe she would not know it.[19] Moreover, the theistic proposition confirmed by this experiential evidence must also be open to disconfirmation;[20] that is, the experience could theoretically confirm a proposition expressing the existence of a supernatural being that lacks the omni-predicates [21] or possesses a feature incompatible with the omni-predicates, whereby this being and an omni-being cannot both exist.[22]

Counter Argument

This is a very strong argument that only really applies to the individual claiming to have a divinely caused uncanny experience. Notwithstanding, how does the experient, by introspection, come to know she is not deluded? Moreover, how can this be tested by her or others? It cannot, and Yandell's argument, though convincing, is hypothetical and the criteria far too demanding.

Dramatic Conversion

From the above, it seems that personal interpretation governs whether an uncanny experience is religious or naturalistic for the experient. A predisposition to either a theistic or a naturalistic explanation would account for such interpretive variation. There are a variety reasons for differing predispositions that generally relate to the life history of the experient.

Unfortunately, an uncanny experience, prior to the interpretation process,[23] is not labelled *religious* or *natural*; consequently, other evidence is required to determine the veracity of the differing interpretations; that is, purely religious or purely naturalistic.[24]

People with differing religious convictions have explanations supporting their interpretations; some theistic ones are outlined above and some naturalistic ones below. However, none are conclusive and the threat of an impasse remains. However, although a conclusive explanation seems beyond reach, there may be evidence that provides theism with a probabilistic advantage; that is, the force of some uncanny experience that results in an unexpected belief-conversion.[25] If there is evidence that some types of uncanny experiences were such as to convert the experient to theism then such evidence, if it exists, could be reasonably persuasive. Moreover, it would be fair to conclude that this type of converting uncanny experience is more liable to occur if God exists rather than not.[26]

Counter Argument

Such evidence is not conclusive as there could be alternative explanations other than God's will; nevertheless, the argument is persuasive. However, it is one thing postulating dramatic converting evidence, it is another finding it. Indeed, the important issue for the Bayesian is whether such evidence exists; this is essential for the conditionalization process. Given that the inclination of the atheist or agnostic is to reject or be skeptical of an uncanny experience as being caused by God, then I doubt there will be many cases of dramatic conversion episodes. I will return to the issue of evidence under *Bayesian conditionalization*.

CONFIRMING THE EVIDENCE ON THE NEGATION OF THE PROPOSITION

I will now consider arguments that attempt to tilt the balance in favor of the atheistic viewpoint; that is $\Pr(e|\neg p \,\&\, k) \geq \Pr(e|p \,\&\, k)$. It is very unlikely that the left-hand function; that is, assuming God does not exist, will be greater than the right-hand function. However, the atheist would win the argument with equality; that is the evidence would exist whether God exists or not. Much depends upon the auxiliary assumptions in *k* and their justification. (See Appendix 1) In the same style as above, I will provide some counter arguments and replies for consideration.

Anti-Religious Experience

In terms of disconfirming evidence of the theistic proposition, the atheist could also cite an *anti-religious* experience of an intense nature; for example, the experience of witnessing the horrendous 1994 Tutsi and Hutu genocide in Rwanda. Indeed, this and other similarly horrendous events, may have converted some theists to atheism; *apostasy*. Also, the *experience* of not experiencing; that is, being aware of others claiming to have religious experiences, but you do not. Given God's love for all mankind, this raises the issues of deception and favoritism which is counter to God's attributes.[27]

Counter Argument

Human and animal suffering is probabilistically accounted for in the previous chapter; the problem of evil. It cannot feature in this chapter as an additional probabilistically disconfirming evidential element in the cumulative Bayesian analysis as this would be double counting.

The absence of a personal experience does not entail the non-existence of the cause of experience; indeed, an uncanny experience could be misinterpreted as naturalistic when it had a divine cause; this may be common and someone's assertion that such experience evades them could be false. Moreover, unlike natural causation, divinely caused experience involves divine intention and its restriction could be for a greater good; that is, beyond our ken.[28]

Reductionism

It could be argued that there is a relationship between non-religious uncanny experiences and ecstatic uncanny experiences claimed to be divinely caused, as the edifying pleasure can be very similar. Indeed, such similarity fuels the sceptic's argument; that is, divine explanations can be reduced to naturalistic explanations like psychosis without reference to religious meaning.

Counter Argument

There are cases that can be explained by reducing the experience to a psychotic episode, drugs, or hysteria, but then to extrapolate these explanations to all uncanny experiences is bad inductive methodology. Just one example that does not conform to the hypothesis contradicts its universal application.[29] Moreover, to select extremes as representing most cases of uncanny experience claimed to be religious is to introduce bias into a naturalistic explanation.

Religiously interpreted uncanny experience has a non-religious input; for example, context, culture, past experiences, and expectation, but such input is not exhaustive in terms of explanation, the religious interpretation also has explanatory relevance; reductionism is always in danger of evisceration. Indeed, naturalistic reductionism in this context can be analytically restrictive, failing to capture the full picture of the experience.[30]

Bias

It could be argued that religiously interpreted uncanny experience is riddled with bias and that those proclaiming it as evidence for the existence of God are ignoring other better explanations for such experiences provided by psychology and physiology. For example, wishful thinking, hallucinations, drugs, mental confusion with aesthetic experiences, hysteria, auto-suggestion, self-delusion, and a host of other possible explanations under the secular sciences of sociology and anthropology. The reason for selecting divine causation over naturalistic causation for the experience is bias produced by something other than the experience; for example, a prior belief in God and a religious world-view. Such prior beliefs are frequently accompanied by behavioral tendencies such as being a regular church goer and deference to a preacher who encourages the quest for connectivity with God. Consequently, rational comparison of all the possible explanations are immediately eschewed in favor of a divine explanation.

Counter Argument

A predisposition, psychological disposition or extrinsic influence may bias interpretation, but having no interpretation whatsoever dilutes the experience; that is, the experience becomes meaningless and without content. Indeed, all the different experiences we have usually trigger an interpretation of some sort. Interpretive balance is attained by reflecting upon the possible different interpretations and then selecting by way of inference to the best explanation.[31]

As I argue above, a naturalistic explanation can also be bias; that is, eschewing religious interpretations based on something other than the experience; that is, a dominant naturalistic world view. The rational process is to consider all explanations and favor the best explanation. However, what counts as the best explanation is itself subject to bias from the theist and atheist alike. Consideration of the evidence, cooperative dialogue, and possible consensus is a truth-orientated process worth adopting when faced with interpretive disagreement. A process I attempt to harness in this book through intersubjective Bayesian probability theory, but even then, interpretive consensus is not certain.

Falsifiability

There is a lack of falsifiability about religious existential claims based upon a religious interpretation of an uncanny experience; for example, God exists. Such objective existential claims based upon uncanny experience are questionable.[32] Indeed, seeming to experience a rat in the kitchen is an existential claim that can be verified or falsified, seeming to experience God is not. Moreover, naturalistic explanations of uncanny experiences, such as an illusion, do not proclaim the existence of anything other than perhaps a magician and that can be verified or falsified. It seems that naturalistic explanations of uncanny experiences are not so susceptible to the charge of un-falsifiable as those of supernatural claims.

Counter Argument

The claim that some uncanny experiences have a religious interpretation is invalid because the interpretation is un-falsifiable may be a claim too far. A hypothesis cannot be testable in isolation, there are bundles of additional hypotheses that are presupposed when considering its truth;[33] auxiliary assumptions. (See Appendix 1) Consider the following likelihood:

$\Pr(e|h \,\&\, k)$, where:

e: God has spoken to me
h: God exists
k: background knowledge with auxiliary assumptions a_1 to a_n embedded within, where:
a_1: God can communicate with persons
a_2: God wishes to communicate with persons

a_3: There are uncanny experiences that do not have a naturalistic explanation
a_n: Additional auxiliary assumptions

This likelihood can be expressed as $Pr(e|h \& a_1 \& a_2 \& a_3 ... \& ... a_n)$ and if any one of the auxiliary assumptions were false then the entire conjunction would be false. The charge of un-falsifiable presupposes that all the conjuncts are unfalsifiable, but a_3 is and hence the charge is invalid.[34]

Comfort and Wish-Fulfilment

Interpreting uncanny experience as religious is a consequence of the need for the psychological comfort provided by a religious interpretation.[35] Sigmund Freud (1927) viewed this as a quelling of anxiety that is both illusionary and delusional.[36] This position views religious interpretation of uncanny experience as non-evidentially based and consequently irrational or neurotic. Additionally, humans strive to make sense of the world and seek explanation for all phenomena and providing uncanny experiences with a religious explanation is intuitively preferable to no explanation.

Counter Argument

Religious thought has not always been a comfortable enterprise, such notions as a deity of death, damnation, and punishment have, at times, pervaded religious reflection. It has involved abstinence, accusation, and retribution; a personal state of terror. Thus, a comprehensive comfort/wish fulfilment theory is required to explain these rather uncomfortable anomalies.[37] Moreover, it is not the case that all interaction with the world elicits a need for explanation; the aesthetic pleasure of music only requires explanation for those who choose to delve into its complexities such as the musician or composer. Likewise, uncanny experience involves an instinctive rather than reflective interpretation, religious or naturalistic, except when further explanation or justification is called for. A comfort/wish fulfilment explanation of a religious interpretation of uncanny experience may have some traction, but is insufficient as a comprehensive theory.

Collective Consciousness

Emile Durkheim (1961) provides a secular explanation for a religious interpretation of uncanny experiences. He introduces the notion of the 'sacred' being symbolic of and corresponding to society. For members of a society, adapting to and engaging with the society is complex and a symbol has emerged to represent it: the sacred. Durkheim sees societies as organisms and persons, albeit metaphorically. Thus, experience of the sacred is for Durkheim, society becoming self-conscious: "If society is something universal in relation to the individual, it is none the less an individuality itself, which has its own personal physiognomy and its idiosyncrasies; it is a particular subject and consequently particularizes whatever it thinks of" [Durkheim, 1961, 493] Thus, there are two consciousnesses, the individual's and the collective's and experience of the sacred is the individual's experience of the collective consciousness; misinterpreted as being caused by God. Experiences of the sacred are heightened in the group context where there are close relations between the individuals in an assembly, with the experience weakening when the assembly dissipates.[38]

Counter Argument

This is an imaginative and captivating explanation, but societies are not analogous to organisms and persons; they have no physiognomy or mind and do not develop in evolutionary ways.[39] Moreover, some of the most intense religiously interpreted uncanny experiences are reported to be solitary.[40] It is one thing possessing a collective awareness or group harmony, which can be ecstatic; for example, in a sports arena, it is quite another to suggest that in that state of harmony an additional single consciousness emerges that causally affects the individual consciousness of the group members. Consciousness supervenes on brain states, it has no causal efficacy, thus a collective consciousness would emerge from multiple brains presumably interconnected by perception; two types of consciousnesses from the same brains. Ants can form a collective through chemical connectivity and the individuals work for the benefit of the whole; synergy. Perhaps humans do something similar by connectivity through perception. However, such interconnectivity is not complex enough to generate a conscious state. The better explanation is that there are individual emergent consciousnesses from individual brains causally affected by individual perceptions of the behavior of other persons and that the notion of a *collective consciousness* is a metaphor.

Anthropomorphism

Stewart Guthrie (1993) argues that all perceptions are an interpretation and consequently multiple interpretations are always possible; selection is therefore a bet on the most veridical outcome.[41] Further, he suggests that animism; that is, the tendency to see living animals in non-living things such as clouds, is an evolved human trait intrinsic to perception.[42] Consequently, our lights are continually scanning the world for living things, particularly human; we are preoccupied with this function and when the perception is an illusion we animate and/or anthropomorphize. Thus, a sudden uncanny experience can be interpreted as a person-to-person communication and anthropomorphized thereafter; "God the Father is speaking to me." Like awareness of an unseen lurking animal or unexpected human communication, the experience is attention catching.

Counter Argument

This is a persuasive argument, but not conclusive. Uncanny experiences can be brought on by deliberate action; for example, meditation. Thus, the element of surprise is absent yet the experience can be just as gripping on every occasion. Moreover, the question of what causes this uncanny experience that is then anthropomorphized is not addressed. The argument merely draws us back to consideration of alternative causes for the experience; one being divine communication. Hence, the argument from anthropomorphism does not eliminate a divinely caused experience from the set of possibilities, but gives us good reason to give due consideration to the alternatives.

EVALUATORS' CONCLUSION

The arguments are indicative of an impasse; the theist giving support to a divine cause for uncanny experience and the atheist a natural cause; that is, the same evidence equally confirms different hypotheses. This is a problem if uncanny experiences are intended to give probabilistic support for the proposition *God exists*.

The agnostic is liable to be agnostic on the matter and therefore track the atheist's position; that is, cases of uncanny experience are just as liable to emerge whether God exists or not. The only argument that may sway all parties to a theistic explanation is that of dramatic conversion experiences (DCE). Given that uncanny experience is being approached as evidence in support of the theistic proposition, then the evidence for consideration should be DCE. Even then, the evaluators do not think that a complete consensus would be reached on consideration of DCE. It is thought that the agnostic would give some, if limited support for the DCE argument and possibly persuade the atheist to form a consensus with them on a low, yet confirmatory λ-value.

Evaluators' Judgement

Let *e* be DCE (dramatic conversion experience)

1. *Atheist's view*: $\Pr(e|p \ \& \ k)$ is greater than $\Pr(e|\neg p \ \& \ k)$
2. *Agnostic's view*: $\Pr(e|p \ \& \ k)$ is greater than $\Pr(e|\neg p \ \& \ k)$
3. *Theist's view*: $\Pr(e|p \ \& \ k)$ is much greater than $\Pr(e|\neg p \ \& \ k)$

Consensus

An expected consensus between the atheist and agnostic as follows:

$\Pr(e|p \& k) = 0.60$
$\Pr(e|\neg p \& k) = 0.40$
$\Pr(e|p \& k)/\Pr(e|\neg p \& k) = \lambda = 1.5$

However, the theist would be unlikely to compromise, preferring the following values:

$\Pr(e|p \& k) = 0.80$
$\Pr(e|\neg p \& k) = 0.20$
$\Pr(e|p \& k)/\Pr(e|\neg p \& k) = \lambda = 4$

It is worth noting that these differences between the members can be the result of different interpretations of the auxiliary assumptions; namely, a different account of what is a justifiable auxiliary assumption and what is not. (See Appendix 1)

Prior Probability

The follow-on priors from evidential element 4 from the previous chapter that apply to this evidential element are atheist: 0.07; agnostic: 0.40; theist 0.79.

Bayesian Conditionalization

Most surveys on religiously interpreted uncanny experiences focus on the nature and quantity of those experiences across different groups identified by the researchers, but there is little emphasis on the atheist and agnostic groups. David Hay and Ann Morisy (1978) undertook such a survey and part of their survey compared religiously interpreted experience across religious groups; for example, the Catholics group and Jewish group. The agnostic, atheist, and don't know groups were included in this survey and when individually asked whether they have had an "awareness of a presence or power" produced the following results:[43]

Table 5.1. Hay, Morisy Results

Groups	*Negative*	*Positive*	*Numbers*
Agnostic	77%	23%	44
Atheist	76%	24%	37
Don't know	77%	23%	35

This table provides a percentage comparison in the response to the "awareness of a presence or power" question between groups comprising agnostics, atheists, and individuals who are unsure of their conviction.

"No doubt many of them would not interpret these experiences as religious, but that there should be such a large proportion of positive responses from them raises interesting questions." [Hay & Morisy, 1978, 259f] This finding repeated itself in David Hay's (1979) survey.[44] Recorded conversion experiences in this survey were small; only four and it is not clear if they were atheists prior to the conversion.[45]

Edwin Starbuck (1897) undertook a survey into religious conversion and although not directly referring to atheism or agnosticism did refer to a segment of the respondents as having a 'tendency to resist conviction' (religious conviction); 41% of the respondents experienced this tendency with all these respondents converting.[46] However, the survey did not correlate this tendency to uncanny experience, although the process of conversion could be described as an uncanny experience with religious connotations.

Bob Atemeyer and Bruce Hunsberger (1997) conducted a survey of approximately 4,264 students in two Canadian universities in 1994 and 1995 seeking information on a variety of religious views, one being dramatic conversion events termed "amazing believer cases"; a rare element of the population; only 0.8%.[47] An amazing-believer case being those members of the population that quickly became Christians from a non-religious background; them and their family background.[48] This could be described as an *awakening*, and was described as such by one interviewee.[49] This raises the question of whether such an *awakening* can be described as an uncanny experience in itself as contrasted to a vision or a voice for example. In most cases, there was some religious connection made at the commencement of the awakening; for example, reading the Bible and or contact with religious friends. Moreover, the conversion resulted in a sense of well-being post-awakening.[50] Because of the speed of conversion from a state of indifference, this could be described as an uncanny experience with religious connotations, what Davis (1999) termed a *regenerative experience*.[51]

The above is not overwhelming evidence to say the least; nevertheless, it is DCE albeit limited in quantity. The theist would probably view it more positively than the atheist, and the agnostic would give it some credence and hold at λ value 1.5. Because of the paucity of evidence, the evaluators believe that the agnostic/atheist consensus would fracture and the atheist would be inclined to reduce their λ-value to 1; that is, $\Pr(e|p \,\&\, k) = \Pr(e|\neg p \,\&\, k)$ and no confirmation of the theistic proposition. The theist would probably be disappointed with the paucity of evidence for DCE and may be persuaded to form a consensus with the agnostic with a λ-value of 1.5.

Posterior Probability

The λ-value of 1 for the atheists ensures no confirmation for this sub-group; thus, their posterior probability value for the truth of the theistic proposition remains unchanged at 0.07.

From Bayes' theorem (likelihood ratio form):

$$\Pr(p|e \,\&\, k) = \frac{\lambda \times \Pr(p|k)}{[\lambda \times \Pr(p|k)] + [1 - \Pr(p|k)]}$$

Substituting the two prior probabilities of the agnostic and theist into the theorem:

Agnostic

$$\Pr(p|e \,\&\, k) = \frac{1.5 \text{ x } 0.40}{[1.5 \times 0.40] + [1 - 0.40]} = 0.50$$

Theist

$$\Pr(p|e \,\&\, k) = \frac{1.5 \text{ x } 0.79}{[1.5 \times 0.79] + [1 - 0.79]} = 0.85$$

§ SECTION TWO: MIRACLES

The danger of circularity lurks in this sixth evidential element in the same way as that of religious experience; that is, if what counts as a miracle includes God as the cause, then miracles cannot be employed to support the existence of God without incurring a charge of begging the question. In fact, like religious experience, a miracle must have religious significance;[52] that is, have a divine or quasi-divine cause.[53] Given this, the atheist is unlikely to predicate an inscrutable event as miraculous unless using the term metaphorically.

Extreme Enigmatic Events

There are enigmatic events of the puzzling kind that occur to everyone; for example, bumping into a friend on holiday on the other side of the world.

However, the enigmatic events that concern this section are of the extreme kind or near extreme kind; for example, raising the dead and sudden inexplicable medical cures.

To escape the problem of begging the question and the charge of bias, I will normally refer to miracles as *extreme enigmatic events presumed to be divinely caused* or something similar and reduce the use of the term miracle to avoid surreptitious bias. Therefore, the evidential element under consideration becomes *extreme enigmatic events* (EEEs).

CONFIRMING THE EVIDENCE ON THE PROPOSITION

I will now consider arguments that attempt to tilt the balance in favor of the theistic viewpoint; that is $Pr(e|p \& k) > Pr(e|\neg p \& k)$. However, most philosophical arguments in this domain are directed towards discounting miracles and most of the arguments presented do just this. Notwithstanding, as with the previous two chapters, I will provide counter arguments and replies for consideration.

Likelihoods

From a probabilistic point of view all three conviction groups would agree that divinely caused EEEs are clearly more likely to occur if God exists than if He does not; that is $Pr(e|p \& k) > Pr(e|\neg p \& k)$. However, some enigmatic events may occur for reasons other than God's will; for example, mysterious medical cures, but if God exists, then there is a greater probability that these events, particularly EEEs such as walking on water and raising the dead, would occur than if He did not exist.

Counter Argument

From a Bayesian conditionalization perspective, there must be actual evidence of divinely caused EEEs occurring for this theoretical perspective to be accepted by all three conviction groups. This raises the question of how we know that God exists to cause them. On this basis, when it comes to considering Bayesian conditionalization, much will depend upon the prior probabilities of the theistic proposition held by all three conviction sub-groups. Their degree of belief in the theistic proposition will affect their judgement on whether some reported EEEs are divinely caused.

As with religious experience, the issue of interpretation returns. The theist may interpret EEEs as being divinely caused, whereas the atheist's interpretation is that they may be naturally caused or more probably, fictitious. Indeed, the atheist, unless converted by an EEE or the report of one, will never accept a religious interpretation of it, as such acceptance would acknowledge the existence of God. If there is evidence of a fair number of atheists witnessing EEEs and consequently converting, then this may be tilting evidence for other atheists to give some credence to the possibility of a divine cause. However, I doubt if there is such evidence as even the theists would acknowledge that miracle-claims are few and far between.

Self-Evident

There is a range of enigmatic events from puzzling events such as thinking of someone and suddenly they telephone you, through to weird events like particle entanglement, through to quite perplexing events such as inexplicable medical cures, through to the classical *miracles* such as the resurrection of Jesus. At the extreme end of this enigma-spectrum no other explanation is possible other than divine causation if the event did occur; it is self-evident that they are divinely caused.

Counter Argument

Although the Biblical miracles cannot generally be explained by the effects of natural law, even probabilistic law, the whole issue of veracity of testimony arises; an issue I address below. Could the witnesses to these events be deceived or are they deceivers or even non-existent? Putative attitude to witness reports is positive, but is balanced by contingent knowledge and understanding. Resurrection, walking on water, turning water into wine and raising the dead are not part of everyday experience. The witness reports are not self-evidently true; belief-dissonance arises because of the incommensurability with world-views of those judging the event.

Perhaps criteria could be agreed by all parties as to what would count as divinely caused EEEs, such as Charles Leslie's (1697) criteria; in summary: the facts are actually perceived by persons; witnessed by many; evoked through monuments and ceremonies and observances conducted from the time the event occurred.[54] However, given the atheist's skepticism, the criteria should be far more discerning; for example, if the event is assumed to be a violation of a natural law this law must not be false. Neither must the report be an invention, a political or religious maneuver, a misinterpretation, illusion, or delusion on behalf of the observer or observers. These are very stringent criteria and I doubt if any reports would survive such criteria in an atheist's lights, as acknowledging that they did would imply a religious interpretation; an anathema to an atheist.

Divine Connectivity

It would be through divinely caused uncanny experiences and EEEs that God makes His presence known to persons at apposite moments in world history and in an unmistakable fashion.[55] In particular, responding to human petition through prayer and in support of prophets.[56] This is particularly pertinent for a God that must, by necessity, remain hidden.[57] Moreover, if a universe that will spawn life must be by necessity essentially probabilistic, then off-plan events will occur by divine intervention to ensure God's ultimate plan.

Counter Argument

This argument assumes that God exists. However, one reason why divinely caused EEEs do not occur, given the existence of God, is that if he instantiated the universe, then He would have created the best of all possible worlds,[58] including the laws of nature therein. If God is omnipotent, He would have no reason to cause such laws to be violated; to do so would be a breach of His omnipotence.

Reply

This counter argument presupposes the divine violation of natural laws, something I argue against below.

CONFIRMING THE EVIDENCE ON THE NEGATION OF THE PROPOSITION

I will now consider arguments that attempt to tilt the balance in favor of the atheistic viewpoint; that is $\Pr(e|\neg p \,\&\, k) \geq \Pr(e|p \,\&\, k)$. *Prima facie*, it is very unlikely that the left-hand function will be greater than the right-hand function, although such a view requires justification. Nevertheless, the atheist would win the argument with equality; that is the evidence would exist whether God exists or not. In the same style as above, I will provide some counter arguments and replies for consideration.

Double Counting

Some might argue that EEEs, thought of as divinely caused, are no more than a type of religious experience and that to consider it as a separate evidential element in support of the theistic hypothesis is double counting.

Counter Argument

This is not the case, uncanny experiences are about a type of mental event undergone by a person or persons, whereas EEEs need not result in such a mental event; it may not even be witnessed. However, if such an event is witnessed, it could result in an uncanny experience; notwithstanding, they are not identical.

Undiscovered Laws

There are inexplicable natural events; for example, the continued expansion of the universe has been put down to dark energy, but no such energy has been discovered yet. Likewise, particle entanglement has been mathematically established with some confirming evidence, but it remains an explanatory enigma. EEEs tend to be viewed as violations or quasi-violations of putative natural laws, but it is possible that there are laws that would explain enigmas that remain undiscovered and account must be taken of this possibility.[59]

Counter Argument

Contrary evidence of a natural law may only slightly reduce the probability of its truth provided the mass of evidence is supportive. To accept the explanation of an unexplained-law rather than a violation of a natural law would beg the question. Adopting the principle that apparent violations of natural laws presuppose unknown natural laws prior to consideration of an apparent violation and forming the conclusion that this apparent violation is an unknown natural law is circular. Much depends on the detail of the enigmatic event, some lend themselves to the possibility of an unknown natural law; particle entanglement, some to the violation of natural law; Jesus walking on water (Matthew 14:22–33).

Hume's Rejection

David Hume [1748] (1993) places five obstacles in the way of any claim that belief in miracles is rational:[60]

1. There are no miracles attested to by sufficient numbers of qualified observers.[61]
2. People are simply drawn to "surprise and wonder."[62]
3. Reports are predominantly from "ignorant and barbarous nations."[63]
4. Different religions frequently contradict each other in miraculous reports.[64]
5. People are inclined to support their chosen faith and are thus deluded.[65]

Counter Argument

Points 1, 2, and 5 have some validity; numbers of credible independent witnesses are important as I argue below, and the most bizarre groups are formed upon questionable claims such as vampires, UFOs, and astrology for example. Moreover, bias or even irrational acceptance of group-tenets is very common in and outside the religious domain. However, points 3 and 4 can be contested; contemporary enigmatic medical cures resulting from prayer are being claimed in some quarters of 'western civilized' countries and there is a greater detente between differing religions and their beliefs. Notwithstanding, Hume's obstacles do provide good reason to treat claims of divinely caused EEEs with caution.[66]

Hume; Laws and Testimony

A claim that an EEE occurred that was a violation of a natural law is evidence that that the claim is liable to be false, as a natural law is a universal claim making the event highly improbable. Indeed, David Hume [1748] (1993) argues that the primary refuting evidence against the rational acceptance of miracles are the laws of nature; that is, our background knowledge of the laws that acts as a verifying lens upon confirming evidence such as testimony. Hume does not deny that miracles may have occurred, but is questioning the conjunction of miracles and natural law violations:[67] "A miracle is a violation of the laws of nature; and as a firm and unalterable experience has established these laws, the proof against a miracle, from the very nature of the fact, is as entire as any argument from experience can possibly be imagined." [Hume [1748], 1993, X, 1, 76] Moreover, testimony in Hume's view, cannot ameliorate such proof: "…no testimony is sufficient to establish a miracle, unless the testimony be of such a kind, that its falsehood would be more miraculous, than the fact, which it endeavors to establish:…" [*Ibid;* X, 1, 76] Here, Hume is cynically arguing that testimony is only able to establish the truth of a miracle if and only if the miracle is already more probable than not; that is, he gives little credence to the force of testimony in the case of miracles.[68]

Counter Argument

Hume's attack upon the power of testimony can be answered to some extent. Indeed, reflecting upon Hume's criticism of the reliability of testimony in support of miracles, Philip Dawid and Donald Gillies (1989) provide a revealing Bayesian analysis. They consider two prior probability values for a proposition expressing the occurrence of a miracle; $\Pr(M|k) = 10^{-6}$ and $\Pr(M|k) = 10^{-3}$; where M is the miracle proposition.[69] They then consider the proposition expressing the testimony T of one or more reliable witnesses of the miracle-event.[70] The analysis proceeds as follows:

1. *From Bayes' theorem:*
   $\Pr(M|T \,\&\, k) = \Pr(T|M \,\&\, k)\Pr(M|k)/\Pr(T|k)$
2. *From the theorem of total probability*:
   $\Pr(T|k) = \Pr(T|M \,\&\, k)\Pr(M|k) + \Pr(T|\neg M \,\&\, k)\Pr(\neg M|k)$
3. *From 1 & 2*:
   $\Pr(M|T \,\&\, k) = \Pr(T|M \,\&\, k)\Pr(M|k)/\Pr(T|M \,\&\, k)\Pr(M|k) + \Pr(T|\neg M \,\&\, k)\Pr(\neg M|k)$

Dawid and Gillies then make the following subjective judgement: $\Pr(T|M \,\&\, k) = 0.99$; $\Pr(T|\neg M \,\&\, k) = 10^{-3}$. If a miracle did occur then we would expect reliable witnesses to testify to that fact, but if it did not occur, the likelihood of such reliable witnesses testifying that it did would be very low indeed. Taking these and the prior probability values above then:

4.1. *From 3*: $\Pr(M|T \,\&\, k) = 0.99 \times 10^{-6}/\ 0.99 \times 10^{-6} + 10^{-3}(1\text{-}10^{-6}) \approx 10^{-3}$
4.2. *From 3*: $\Pr(M|T \,\&\, k) = 0.99 \times 10^{-3}/\ 0.99 \times 10^{-3} + 10^{-3}(1\text{-}10^{-3}) \approx 0.5$

Thus, argue Dawid and Gillies from 4.1 Hume is correct "that no amount of testimony can establish that a miracle has occurred…" [Dawid & Gillies, 1989, 64] However, from 4.2 "…very good testimony might, after all, make it reasonably probable that a miracle had occurred." [*Loc. Cit.*]

It is the selection of the priors that makes the difference and it is background knowledge that affects this selection. A propensity to scientific evidence would support $\Pr(M|k) = 10^{-6}$, whereas propensity to theistic considerations would support $\Pr(M|k) = 10^{-3}$.

Credulity and testimony are the essence of witness accounts and we rely on them constantly in society. It is when such accounts conflict with our background knowledge that doubt emerges. This is Hume's point; our knowledge of the laws of nature create such doubt in the case of EEEs. However, Richard Swinburne (2004) challenges Hume, arguing that background knowledge can also include probabilistic evidence of the existence of an omnipotent being from evidential elements other than EEEs. Such a being could cause a violation of a natural law.[71] Thus, much depends upon the correspondence between testimony and background knowledge and the difference between an atheist's perspective and a theist's is paramount.[72]

Is it the case that divinely caused EEEs are violations of natural laws? Richard Swinburne (2004) believes they are; he views such events as violations (or quasi-violations; see below) of a law of nature by the action or permission of God; that is, a violation of the powers and liabilities of substances[73] by "some uncaused event of a kind that will not be repeated in similar circumstances; or if some cause from outside the system of law-governed objects intervenes to bring about an event not permitted by laws." [Swinburne, 2004, 278]

The laws of nature echo the uniformity of nature and are determined by calculation, experimentation, and observation. They apply at both the macro and micro levels; however, at the micro level such laws take on a probabilistic nature. Given that all matter comprises particles that are subject to quantum indeterminacy, then the notion of *violation* is questionable. Unusual events at the macro level due to probabilistic events at the micro level are possible, albeit very unlikely. Swinburne terms the occurrence of unlikely events due to quantum indeterminacy as *quasi-violations*.[74] Thus, such an improbable event at the macro level would not necessarily breach any natural laws. However, divinely caused EEEs are events involving an intentional divine cause rather than a probabilistic blip of nature.

Laws of nature have application within a closed system and events occur within the system that are governed by these laws. However, an extrinsic cause entering the system could violate these laws if such laws are deterministic. However, given that the laws of our universe are fundamentally probabilistic, then an EEE at the macro level due to chance could occur. Thus, it is possible within the closed system, although extremely improbable, that a person could walk on water at a particular moment. Any extrinsic cause entering the closed system could affect the probabilistic state of affairs so that the EEE occurred. Thus, should Divine Will enter the system and intentionally harness the probabilistic nature of the system then, the EEE could occur. This is not a violation or even quasi-violation of natural law, as the EEE could occur without intervention, rather it is a harnessing of laws so the event occurs at a particular moment. Thus, "…we can give a coherent definition of a miracle as a supernatural intrusion into the normally closed system that works in accordance with those laws…" [Mackie, 1982, 22]

A further Bayesian approach also supports the miracle argument based upon evidential detail rather than testimony as evidence. Actual evidence could be; for example, Jesus walking on water.

Let…
M be the proposition 'miracles have occurred.'
E be the evidence 'Jesus walked on water.' (Matthew 14:22–33)

1. From Bayes' theorem (comparative or odds form, without k for simplicity):
$Pr(M|E)/Pr(\neg M|E) = Pr(E|M)/Pr(E|\neg M) \times Pr(M)/Pr(\neg M)$

2. *Assumption*: $Pr(E|M) \approx Pr(\neg M)$. This is based upon both functions having high probability values; namely, close to 1. If miracles have occurred, as is assumed in the left likelihood function, then given Biblical background knowledge in k (not shown) the evidence E on M is high. Likewise, the notion of a miracle is a very unusual and rare event; thus, the probability of M prior to any supportive evidence is low and by contrast the probability of the negation of M prior to any supportive evidence is high.
3. *From 1 & 2*: $Pr(M|E)/Pr(\neg M|E) = Pr(M)/Pr(E|\neg M)$
4. *Assumption*: $Pr(M) > Pr(E|\neg M)$. This is based on the argument that, given the assumption that miracles have never occurred as assumed in the right-hand function, the evidence E is extremely improbable; far more improbable that is than the prior probability of M, although both functions have a low value.
5. *Conclusion from 3 & 4*: Given that $Pr(M) > Pr(E|\neg M)$ then, $Pr(M|E) > Pr(\neg M|E)$. Also, given that $Pr(M|E) + Pr(\neg M|E) = 1$ then, $Pr(M|E) > 0.5$

Of course, the above is only reasonable in principle and provides a response to Hume, but Bayesian conditionalization requires that the evidence is actual and not just theoretical, which brings the argument back to the veracity of the evidence.

History

Anthony Few (1968), reflecting upon Hume's arguments, provides an argument based upon the methodology employed in historical reasoning. He argues that interpretation of historical evidence presupposes fundamental regularities in the time-line from the past to present. Moreover, such interpretation also requires the application of what counts as possible and probable. However, EEEs thought by some to be divinely caused, suggest a violation of regularity and fail to meet the historian's notion of what is possible and or probable; these failures preclude reports of such events as veridical.[75]

Counter Argument

Events of this type are not considered as regular events in nature, but rare, enigmatic occurrences; therefore, the acceptance of them falls outside that of conventional historicity. Moreover, above I argue against the notion of violations of natural law.

EVALUATORS' CONCLUSIONS

As the argument in favor of miracles is heavily influenced by historical testimony, prior religious conviction as background knowledge plays a crucial role in judging whether an EEE was divinely caused. Moreover, EEEs suggest divine causation, which is generally incommensurable with a contemporary world-view giving rise to questioning the reliability of witness testimony. Again, the religious conviction of those judging whether there was divine involvement or whether the testimony is fictitious is central. The prior-values of the theistic proposition by members of the syndicate play a crucial role in judging whether EEEs are divinely caused; that is, are miracles.

Evaluators' Judgement

To remind the readers that e is not evidence of miracles, but evidence of extreme enigmatic events (EEEs), some of which are thought to be naturalistic; for example, particle entanglement and some of which are thought, by some, to be divinely caused; for example, turning water into wine.

1. *Atheist's, agnostic's, and theist's views*: $\Pr(e|p \,\&\, k)$ is greater than $\Pr(e|\neg p \,\&\, k)$. This is because an EEE is more likely to occur if God exists than if He does not and the former function assumes God's existence. However, there are naturalistic EEEs which await a scientific explanation and these events reside in the background knowledge k of both functions. However, Bayesian conditionalization on evidence is questionable for the atheist as they are unlikely to accept a divinely caused interpretation of an EEE.

Consensus

The evaluators believe that all three hypothetical members would reach a consensus as follows:

$\Pr(e|p \,\&\, k) = 0.80$
$\Pr(e|\neg p \,\&\, k) = 0.50$
$\Pr(e|p \,\&\, k)/\Pr(e|\neg p \,\&\, k) = \lambda = 1.6$

Prior Probability

The follow-on priors from evidential element 5 from the previous chapter that apply to this evidential element are as follows atheist: 0.07; agnostic: 0.50; theist 0.85.

Bayesian Conditionalization

Miracle reports are common in most if not all religions; in the case of Christianity they are well-recorded.[76] There are enigmatic events that are less extreme as the Biblical miracles; for example, healing through prayer and religious sites.[77] Interestingly, belief in miracles in Britain is fairly strong (29% in 2008) and in the USA, very strong (79% in 2010).[78]

There are EEEs where a *full* naturalistic explanation seems to be more appropriate but not yet forthcoming; for example, dark matter, dark energy, quantum entanglement, the Big Bang, and others.

Given this historical and recently witnessed evidence it is reasonable to assume conditionalization by the agnostics and theists, particularly as their theistic priors are robust at this point in the cumulative process. Thus, their interpretation of the evidence would accommodate the notion of a divine cause. However, the atheist's theistic prior is far from robust and their interpretive attitude would be set against accepting this evidence as divinely caused; they are unlikely to conditionalize on it so their posterior probability value for the truth of the theistic proposition on the evidence remains unchanged.

Posterior Probability

From Bayes' theorem (likelihood ratio form):

$$\Pr(p|e \,\&\, k) = \frac{\lambda \times \Pr(p|k)}{[\lambda \times \Pr(p|k)] + [1 - \Pr(p|k)]}$$

Substituting the two prior probabilities of the agnostics and theists into the theorem:

Agnostic

$$\Pr(p|e \,\&\, k) = \frac{1.6 \times 0.50}{[1.6 \times 0.50] + [1 - 0.50]} = 0.62$$

Theist

$$\Pr(p|e \,\&\, k) = \frac{1.6 \times 0.85}{[1.6 \times 0.85] + [1 - 0.85]} = 0.90$$

§ SECTION THREE: THE *CONSENSUS GENTIUM* ARGUMENT

The *consensus gentium* argument is the seventh evidential element for consideration and, in contemporary terms, states that if most people in the world believe in the existence of God, this probabilistically confirms God's existence. The argument is one that Cicero invokes for belief in the existence of the gods in his De Natura Deorum (45BC).[79]

CONFIRMING THE EVIDENCE ON THE PROPOSITION

I will now consider arguments that attempt to tilt the balance in favor of the theistic viewpoint; that is, $\Pr(e|p \,\&\, k) > \Pr(e|\neg p \,\&\, k)$.

Aggregation

Loren Meierding (1998) presents a Bayesian argument in support of the theistic proposition on the evidence that the majority of people believe that God exists and the argument takes the form of the likelihood ratio method I adopt herein. He provides two alternatives for the theistic proposition:

1. An omnipotent, omniscient, and omni-benevolent personal being exists (classical theism) and
2. An omnipotent and omniscient personal being exists (non-classical theism)

I only address 1, as it is only classical theism that I am concerned with herein; that is, a being possessing the above omni-predicates.[80] He also provides two alternatives for the falsity of the theistic proposition:

3. "…Supernatural personal beings with limited power or knowledge exist." and
4. "…no supernatural personal beings exist." [Meierding, 1998, 275f]

Again, for the same reason, I only address 4 in the form of the negation of classical theism.

Consider the likelihood $\Pr(e|\neg p \,\&\, k)$, where e means the majority of people believe that God exists, $\neg p$ the negation of the theistic proposition and k background knowledge.[81] This likelihood function defines the probability that the majority of people will believe that God exists assuming God does not exist; that is a naturalistic causation of the belief-majority.[82] Naturalistic factors (residing in background knowledge k) that result in the belief-majority e would still take effect whether an omni-being exists (p) or does not exist ($\neg p$).

Where God is assumed to exist in the case of the likelihood $\Pr(e|p \,\&\, k)$, then we would expect the majority-belief e to be the case because of the causal potential of the omni-predicates. This is particularly so with the attribute of omni-benevolence; it is expected that God would desire a two-way relationship with people, perhaps through "prayer and implanted beliefs," and given the other two attributes God has the wherewithal to promote this desire.[83] Hence, combining the naturalistic causation of *e* with the divine causation of *e* would result in an increase in the number of believers; an aggregation argument. There is the possibility of over-determination; that is, naturalistic causation is not necessarily added to by divine causation with a proportion of believers; either cause on their own would suffice. On the other hand, there will be a proportion that succumb to belief because of the combined effect and there will be yet another proportion that only succumb to one or the other of the two causal factors. From this, there will be an overall increase given this combination.

From the above we can conclude that $Pr(e|p \& k) > Pr(e|\neg p \& k)$ on the grounds that a combined causal force; that is, expectation concerning God's will and naturalistic factors within *k* is greater than just a naturalistic one and hence, the probability of *e* is greater for the left-hand function than the right.

Counter Argument

In my view, Meierding is too bullish in the force he gives to this argument, particularly if we take into account the three conviction groups that would attempt to reach a consensus on the matter.[84] In fact, naturalistic causation on its own may be strong enough to result in a majority-belief (50% + 1) irrespective of divine causation and even if, in combination, numbers would be greater and hence more feasible that belief-majority *e* is true, if our background knowledge within *k* is such that naturalistic causes are potent enough to get a majority, then $Pr(e|p \& k) \approx Pr(e|\neg p \& k)$. It matters not if there would be a 90% belief-majority in a natural/divine combination and a 51% belief-majority for a naturalistic cause only, as both have passed the mark.

Reply

The very fact of greater potency in the combination state of affairs means there is a greater chance of passing the majority mark. Nevertheless, if the naturalistic forces are considered potent the likelihood ratio $Pr(e|p \& k)/Pr(e|\neg p \& k)$ diminishes. Consequently, much depends on whether it is thought that naturalistic causes can tip the belief/non-belief balance over 50% in favor of belief.

Success Through Truth

Meierding believes that $Pr(e|\neg p \& k)$ is low, arguing that true beliefs are more liable to flourish than false ones as a consequence of the success true beliefs bring; false beliefs are "winnowed out" over time. This is the case at the individual, group and societal levels he argues. Meierding argues that this also has application to beliefs concerning deities.[85]

Counter Argument

This argument does not hold, as only certain types of true beliefs are evidentially beneficial; for example, believing tigers are dangerous. False beliefs such as Zeus resides on Mount Olympus would have little benefit or harm to the believers. Moreover, the tendency to hold false beliefs can be beneficial; for example, seeing a lurking bear rather than a gnarled tree stump keeps the wary traveler on their toes regarding predators as Stewart Guthrie's (1993) argument from animism suggests (see above). Meierding's assertion would need evidence and, pace Meierding, we would expect theistic belief to be on the wane rather than growing because of the emergence of true scientific belief.

CONFIRMING THE EVIDENCE ON THE NEGATION OF THE PROPOSITION

I will now consider arguments that attempt to tilt the balance in favor of the negation of theistic viewpoint; that is $Pr(e|\neg p \ \& \ k) \geq Pr(e|p \ \& \ k)$. Even equality of the functions would be enough to weaken the *consensus gentium* argument.

Functionalism

It may be the case that belief in God could provide a beneficial function irrespective of the truth of God's existence. For example, social cohesion, identity, purpose, and altruism that lead to successful societies and the consequential establishment of religious beliefs. Moreover, wish-fulfillment and comfort explanations can provide a non-theistic account of a belief-majority.[86] It may well be the case that some find comfort and security through their religion in this way; escape from the world of suffering into an imaginary world of comfort and protection. This imaginary world includes the afterlife where injustice is posthumously redressed.[87]

Counter Argument

Certain animal species form cohesive and altruistic social groups without the apparent need for a religious backdrop. Moreover, the prolific polytheistic religions of history have gods whose apparent behavior is contrary to a moral, cohesive, and altruistic society. If society were modeled on their example, then immorality and brutality rather than social cohesion and altruism would prevail.

Functionalist theories presuppose a societal tendency for a false belief in a deity and refer to the prolific spread of such belief, before embarking on alternative reasons for its occurrence; alternative that is to a divine explanation. This creates circularity in the function $\Pr(e|\neg p \,\&\, k)$ when functionalist theories are held within k. It is the question of old evidence[88] and, sitting within k as it does, it is assumed true. If such old evidence presupposes a majority belief in God, then $\Pr(e|\neg p \,\&\, k) = 1$. The method of engaging with likelihood functions when faced with the old evidence issue is to assume its non-existence to tease out the probabilistic force, in this case, of $\neg p$ upon e. Moreover, functionalist theories holding such presuppositions assume p is false, which contradicts the assumed truth of p in $\Pr(e|p \,\&\, k)$ as the functionalist theory resides within k.[89] Thus, caution must be taken with the application of functionalist theories with likelihood ratios; the theories must not specify a false belief-majority or the non-existence of God. Therefore, as a challenge to the *consensus gentium* argument, it does not seem possible that functionalist theories can ensure the truth of a false belief-majority as it is unmeasurable. However, it can be argued that they have some impact on such belief, which is an integral part of this *consensus gentium* argument.

Regarding wish fulfillment and comfort explanations, not all religious doctrines portray the afterlife as Elysium; for example, Hell. Moreover, mental escape from the harsh realities of the corporeal world to the transcendent world of illusionary bliss has been and is often quite the opposite; a deity of death, damnation, and punishment; a religion of abstinence, accusation, and retribution; a personal state of terror, repentance, and regret. Wish fulfillment and comfort theories would be obliged to explain these uncomfortable anomalies.

It is worth mentioning Durkheim's social solidarity theory of collective consciousness as a further naturalistic explanation. I addressed this approach and counter argument under Religious Experience above.

Anthropomorphism

Not addressed by Meierding, but Stewart Guthrie's (1993) argument from the survival benefits of animism and anthropomorphism above is an alternative naturalistic explanation. I have mentioned animism above, but also natural events can be falsely conceived of as being divinely caused because of a natural tendency to anthropomorphize; for example, Thor the God of thunder.[90]

Counter Argument

The false belief argument from anthropomorphism is a two-edged sword. It could just as easily be the case that God harnesses the persuasive force of anthropomorphism to strengthen belief in His existence: "…speaking of God in anthropomorphic terms has been explicitly licensed and welcomed by God." [Griffith-Dickson, 2005, 111]

EVALUATORS' CONCLUSIONS

The *consensus gentium* argument is a strong argument in support of the theistic proposition, but the alternative naturalistic arguments also have some force. However, the argument is a conjunction of both positions and in combination there is a probabilistic advantage over naturalistic explanations on their own. Notwithstanding, there is a possibility of attaining a belief-majority with just the naturalistic explanation.

Evaluators' Judgement

1. *Atheist's and agnostic's view*: $\Pr(e|p \,\&\, k)$ is greater than $\Pr(e|\neg p \,\&\, k)$.
2. *Theist's view*: $\Pr(e|p \,\&\, k)$ is much greater than $\Pr(e|\neg p \,\&\, k)$.

Consensus

The evaluators believe that all three hypothetical members would reach a consensus as follows:

$\Pr(e|p \,\&\, k) = 0.90$
$\Pr(e|\neg p \,\&\, k) = 0.60$
$\Pr(e|p \,\&\, k)/\Pr(e|\neg p \,\&\, k) = \lambda = 1.5$

Prior Probability

The priors from evidential element 6 are: atheist: 0.07; agnostic: 0.62; theist 0.90.

Bayesian Conditionalization

A particularly revealing survey was undertaken through an online panel system and published by the Ipsos Social Research Institute for Reuters News (2011). More than 18,000 people fully participated in the London-based poll in 23 countries.[91] From the countries surveyed 51% believe there is a divine entity, 18% did not, and 30% were not sure.[92]

The interesting find is the 'not sure' category as belief is a variable epistemic state of mind and 'not sure' would represent the agnostic position which could give some additional weight to the believers as well as the non-believers.[93] Although not every country in the world was surveyed it does provide evidence that the majority of people believe that God exists. One problematic country surveyed was China, as for many years the Communist Government tried to eradicate religious belief and any survey is liable to be skewed by this. The sociologist Ch'ing-k'un Yang (1970) provides some insight into religion in China.[94]

There are many individual country surveys that indicate a belief-majority [95] and given the Ipsos survey there is good reason for Bayesian conditionalization to occur.

Posterior Probability

From Bayes' theorem (likelihood ratio form)

$$\Pr(p|e \,\&\, k) = \frac{\lambda \times \Pr(p|k)}{[\lambda \times \Pr(p|k)] + [1 - \Pr(p|k)]}$$

Substituting the prior probabilities of the atheist, agnostic, and theist into the theorem:

Atheist

$$\Pr(p|e \,\&\, k) = \frac{1.5 \times 0.07}{[1.5 \times 0.07] + [1 - 0.07]} = 0.10$$

Agnostic

$$\Pr(p|e \,\&\, k) = \frac{1.5 \times 0.62}{[1.5 \times 0.62] + [1 - 0.62]} = 0.71$$

Theist

$$\Pr(p|e \,\&\, k) = \frac{1.5 \times 0.90}{[1.5 \times 0.90] + [1 - 0.90]} = 0.93$$

This concludes the evidential set and the final probabilistic position of the hypothetical individuals in syndicate. It is worth noting that the atheist has become even more atheistic in the face of all the evidence for and against the existence of God. However, the agnostic has moved into the theistic zone of the belief continuum and hereinafter, to reflect this change, I refer to her as Theist (Agnostic). The theist has moved closer to the point of certainty.

There are other evidential elements that other apologists have considered; for example, consciousness and morality. However, I do not consider them to have any confirming or disconfirming force on the theistic proposition whatsoever. Moreover, it is worth noting that they are both difficult concepts to explain and employing them to support the theistic proposition suggests a 'God of the gaps' approach.

NOTES

1. A *short head* refers to the short distance a horse may win by; i.e., no more than the horse's head length. A short-nose is even less.

2. An analogy being the duck-rabbit drawing whereby it is either seen as a duck or a rabbit. The facts are the same, but the interpretation variable; this is John Hick's *experiencing-as* argument intended to provide a reason for apparent diverse religious experiences, it also weakens the evidential argument from religious experience: Hick J. (1968)—*Religious Faith as Experiencing-As*—Royal Institute of Philosophy Lectures; 2; March 1968; pp. 20–35 and Hick J. (1967); Ch. 5.

3. The Theodicy of Divine Withdrawal; Ch. 4.

4. Alston terms them M-beliefs; M referring to God *manifesting* Himself through religious experience: Alston W. (1993); p. 1.

5. *Doxa* is Greek for belief or opinion.

6. Alston W. (1993); p. 153f.

7. *Ibid;* Chts. 4 & 5.

8. Alston distinguishes between 'rebutters,' 'overriders,' and 'underminers.' Rebutters being that there is good reason to consider a belief as false, overriders being reason to consider adopting a particular belief over another insufficient, and underminers being that there is insufficient reason for adopting a particular belief per se. Alston W. (1993); p. 72.

9. Gellman J. I. (1997); pp. 102–109 and compare Flew A. (1968); p. 126ff.

10. *Ibid*; 6.6; p. 126f and Davis C. F. (1999); p. 123 and p. 174f.

11. The Alister Hardy Society for the Study of Religious Experience continues Hardy's work in this domain.

12. Foundationalism is a theory of knowledge resting on the notion of basic, indubitable beliefs upon which further propositions can be derived to provide a network of known truths. Sense experience being a source of such basic beliefs from which truth about the external world is derived.

13. Basic or properly basic beliefs are not adopted on the basis of other beliefs, but are rationally adopted on face value; for example, I have a pain. Adopted beliefs that are not properly basic are only rational if they can evidentially trace back to properly basic beliefs: Plantinga A (1981); p. 94.

14. *Ibid*; p. 104f.

15. Swinburne R. (2004); pp. 303–322.
16. *Ibid*; pp. 322–324.
17. Gellman J. I. (1997); p. 46ff.
18. *Ibid*; p. 52ff.
19. Yandell K. (1994); p. 274.
20. This echoes the principle of falsification in scientific research.
21. To remind the reader, by omni-predicates I mean *...is omnipotent, omniscient and omni-benevolent*.
22. Yandell refers to these disconfirming features as *collegial* and *lateral* disconfirmation respectively: Yandell K. (1994); pp. 245–149.
23. It has been argued that the very fact of possessing a theistic world view will make the theist's experience different to that of the atheist's rather than the same with a different interpretation. I disagree, for if this were the case, then there would never be conversions from atheism based upon an uncanny experience: Davis C. F. (1999); p. 85 and Alston W. P. (1993); p. 37f.
24. There are variations in interpretation of types of religious experiences; for example, within different religious belief cultures. Such variation does not entail that one interpretation is right and the rest wrong; different interpretations can all be true: Griffith-Dickson G. (2005); p. 402f.
25. James W. (1985); Lecture IX.
26. There are, of course, dramatic cases of apostasy, the causes of which can varied:. Hunsberger B. and Brown L. B. (1984)—*Religious Socialization, Apostasy, and the Impact of Family Background—in Journal for the Scientific Study of Religion* 23; No. 3 (Sept, 1984); pp. 239–251.
27. Griffith-Dickson G. (2005); p. 416f.
28. *Ibid*; p. 417ff.
29. The Inductive Argument; Ch. 4.
30. Griffith-Dickson G. (2005); pp. 393–398.
31. *Ibid*; p. 401ff.
32. Flew A. (1968); p. 129f.
33. Gillies D. (1993); pp. 112ff.
34. Sober E. (2008); pp. 129–131.
35. Guthrie S. (1993); pp. 10–16.
36. Freud S. (1927); p. 31; also Hume D. [1779] (1998); p. 9.
37. Guthrie S. (1993); pp. 72–75.
38. Durkheim E. (1961); p. 386f.
39. Guthrie S. (1993); p. 19.
40. James W. (1985); p. 31.
41. Guthrie S. (1993); p. 42.
42. *Ibid*; p. 45ff.
43. Hay D. and Morisy A. (1978); p. 260.
44. Hay D. (1979); p. 179f.
45. Hay D. (1982); p. 142.
46. Starbuck E. D. (1897); p. 285.
47. Atemeyer R. and Hunsberger B. (1997); p. 30.

48. *Ibid*; p. 231f.
49. *Ibid*; p. 193.
50. *Ibid*; p. 200ff.
51. Davis C. F. (1999); p. 44ff.
52. Swinburne R. (1970); pp. 7–10.
53. By quasi-divine cause I mean caused by someone other than God but empowered to do so by God.
54. Leslie C.[1697] (1815)—*A Short and Easy Method with the Deists*—London: Rivington and Co; p. 13.
55. Paley W. [1794] (1871); p. 3.
56. Swinburne R. (2004); p. 284ff.
57. The Theodicy of Divine Withdrawal; Ch. 4.
58. Gottfried Leibniz first coined the term *best of all possible worlds*: Leibniz G. W. [1710] (1951); p. 128.
59. Coleman D. (1988); p. 338f, Swinburne R. (2004); p. 279 and Everitt N. (2004); p. 123f.
60. I use the term 'miracle' in reviewing Hume's analysis to remain close to the original text, but with the caveat that a miracle is a religious interpretation of an EEE.
61. Hume D. [1748] (1993); X; II; p. 78.
62. *Op. cit.*
63. *Ibid*, p. 79ff.
64. *Ibid*; p. 81ff.
65. *Ibid*; p. 86ff.
66. Mackie J. L. (1982); pp. 14–16.
67. Everitt N. (2004); p. 117f.
68. Gillies provides a Bayesian analysis proving this particular statement by Hume: Gillies D. (1991); pp. 255f.
69. Both prior probability values are very low, but the authors are comparing lottery chances with miracle-probability; hence their choice of low values. Dawid P. and Gillies D. (1989); p. 59ff.
70. M being A and T being a in their paper; *Ibid*; p. 60.
71. Swinburne R. (2004); p. 284.
72. Mackie J. L. (1982); p. 26f.
73. By substances-powers-and-liabilities Swinburne refers to the material that has causal powers to effect change on other substances under the constraint of its liabilities, i.e., under certain conditions such as the causal influence of other substances: Swinburne R. (2004); p. 33f.
74. *Ibid*; p. 281.
75. Few A. (1968); p. 146.
76. Cruz J. C. (1997)—*Mysteries, Marvels and Miracles in the Lives of the Saints*—USA: Tan Books and Howick E. K. (2003)—*The Miracles of Jesus the Messiah*—USA: WindRiver Publishing [First published 1985].
77. Glynn P. (1999)—*Healing Fire of Christ: Reflections on Modern Miracles-Knock, Lourdes, Fatima*—Aus.: Marist Fathers Books.
78. NatCen (2008); Pew Survey (2010).

79. Cicero M. T. [45BC] (1896); II; IV; p. 81f.

80. I argue that non-classical theism is not possible: Omni-Benevolence; Ch. 1.

81. Meierding uses different signifying letters; M for *e*, H and -H for *p*, and ¬*p*, respectively and E for *k*. For purposes of overall consistency, I employ *e*, *p*, ¬*p*, and *k*.

82. Meierding L. (1998); p. 277.

83. *Ibid*; p. 284.

84. *Ibid*; p. 288.

85. *Ibid*; p. 280.

86. Hume D. [1757] (2010); p. 9.

87. Freud S. (1927); p. 31.

88. Old Evidence; Ch. 2.

89. Meierding L. (1998); p. 281f.

90. Hume D. [1757] (2010); p. 8.

91. Argentina, Australia, Belgium, Brazil, Canada, China, France, Germany, UK, Hungary, India, Indonesia, Italy, Japan, Mexico, Poland, Russia, South Africa, South Korea, Spain, Sweden, Turkey, USA. Also, Saudi Arabia partially participated, only answering the question concerning evolution.

92. These percentages add to 99% because of rounding errors.

93. Ipsos Social Research Institute (2011).

94. Yang C. K. (1970)—*Religion in Chinese Society: A Study of Contemporary Social Functions of Religion and Some of Their Historical Factors*—USA: University of California Press [1st published 1961].

95. *Europe: % Belief in God 1999*: Ireland 93; Italy 88; Portugal 93; Spain 81; Austria 83; Belgium 66; France 56; Germany (W) 66; Netherlands 60; UK 61; Denmark 62; Sweden 47: Lambert Y. (2004); p. 31.

Part III

THE WINNING POST

The Chances of Human Resurrection

Before considering human resurrection, it is incumbent to ascertain whether it makes sense to even talk about the resurrection of a dead person. This is even more pressing given the materialist ontology that I have adopted. Thus, upon death the human body decomposes and this decomposition includes the brain; the individual returns to dust and ostensibly ceases to exist. Given this materialist perspective, the individual's mind also ceases to exist and there is no soul that gentle flutters off to Elysium. Notwithstanding, I intend to argue that it is both possible and probable that the dying individual will be resurrected immediately upon death.

This part, comprising the final three chapters of the book, presents my case, with Chapter 6 engaging with the notion of human identity as resurrection does presuppose preservation of identity after death; an important auxiliary assumption. (See Appendix 1) Any resurrection hypothesis assumes it is a possibility and in Chapter 7 I consider this and present, what I believe, is a justifiable possibility-model; one that would answer the charge of impossibility. Possibility is different to probability and in Chapter 8 I given consideration to the probability that, based upon evidence, human resurrection is a reality.

Chapter 6

A Dark Horse

Person-Identity

Consider the state of affairs whereby a person, say Smith, suffers severe Alzheimer's disease, dies in hospital and is immediately resurrected in a parallel world. Also, assume when Smith was alive she did not possess a ghostly immortal soul which persisted after her death. Yet suddenly Smith finds herself in a different world; perhaps she knew she was dying in hospital. Now she is astounded to find herself physically in pristine condition surrounded by other resurrected people, some of whom she recognizes as dead relatives, others she does not; all, like her, are in pristine condition. She pinches her arm and asks herself and those surrounding her "is this really me"? This is the scenario that I am presenting in my argument for human resurrection, but apart from the mechanism of how such a state of affairs is instantiated, several perplexing questions arise. In this scenario, Smith has no reason to think this new person is not her; it seems that her personal identity remains intact. Her resurrected relatives identify Smith as the same person who they once knew when she was alive the last time they met. They embrace her, talk about old times and what she will come to discover in this new exciting world. What does it take for Smith to intuitively comprehend that the resurrected being is in fact her and for the others around her also to believe Smith is their relative? Also, how can Smith's identity persist from one world to the next after her old body and brain died? After all, there are two bodies in the frame, the body prior to death, which has now been cremated and the body after death which is alive and well. They are not the same body, although they share a visual likeness enabling recognition. Thus, the first body *was* Smith and the second body *is* Smith and identity persisted across this corporeal and environmental change; one person and two bodies albeit at different times. These questions and others relate to the problem of person-identity.[1] But the first question to ask is what is meant by identity per se?

Identity

It is important to distinguish between numeric identity and qualitative identity. The former refers to identity in terms of one rather than two things; for example, the morning and evening stars are identical because they are one thing; Venus. The latter defines two things that possess identical properties; for example, identical motor cars. Person-identity is a case of the former; that is, a person as a child is the same as a person as an adult even though they differ qualitatively. It is also important to distinguish between synchronic identity and diachronic identity; for example, what makes my motor car the same as my neighbor's is a case of synchronic identity. What makes Pip, my nine-year-old cat the same as Pip my newborn kitten is a case of diachronic identity.

Identity is formally defined by the indiscernibility of identicals and reflexivity; in turn, these both entail symmetry in identity and transitivity. Thus:

1. *Indiscernibility of Identicals*: this states that if something x is identical to something y, then *all* non-relational properties exhibited by x are exhibited by y. For example, if my car is identical to my wife's car, then all the properties of my car are such just in case all the properties of my wife's car are the same. The fact that we share one car is not relevant to the principle, although it is a consequence. The principle of the Indiscernibility of Identicals is formally presented as $x = y \rightarrow \forall F(Fx \leftrightarrow Fy)$ (where F are non-relational properties[2]).
   1.1. *Identity of Indiscernibles*: This is the converse of the above; that is, instead of defining what logical state of affairs results from two things being identical (above), it is a case of what logical state of affairs ensure two things are identical. The Identity of Indiscernibles is formally presented as:
   $\forall F(Fx \leftrightarrow Fy) \rightarrow x = y$.

Clearly, 1 and 1.1 are associated and also associated with Wilhelm Gottfried Leibniz[1696] (1973); in conjunction they are termed Leibniz's Law.[3]

2. *Reflexivity*: this law is intuitively obvious; every x is identical with itself; presented formally as: $\forall x(x = x)$.

The indiscernibility of identicals and reflexivity entail symmetry and transitivity:

3. *Symmetry*: $\forall(x)\forall(y)((x = y) \rightarrow (y = x))$ For example, if my car is identical to my wife's car, then my wife's car is identical to my car.

4. *Transitivity*: $\forall(x)\forall(y)\forall(z)(((x = y) \& (y = z)) \rightarrow (x = z))$. For example, if Venus is the Morning Star and Venus is the Evening Star then, the Morning Star is the Evening Star.[4]

These four principles should guide our thinking concerning person-identity, but there are other relevant factors.

Persistence

The issue of diachronic person-identity is one of persistence over time and is particularly pertinent to a resurrection scenario in so far as person-identity is claimed to remain intact despite death, time change, body change, and world change. Any reasonable explanation of such a bizarre state of affairs would at least require a set of necessary and sufficient conditions for diachronic person-identity. Thus, is New-Smith the same as Old-Smith? Old-Smith was ill, older in appearance, and mentally impaired, whereas New-Smith is well, young, fit, and mentally sharp. Consequently, Old-Smith and New-Smith are different mentally and physically, but both the resurrected Smith and her resurrected relatives believe she is one and the same person. It could be claimed that Old-Smith and New-Smith are *just not the same person* because of these physical and mental differences. However, used in this sense it is not a claim of failure in persistence of self, but only a claim that Smith has changed for the better, not that she has ceased to exist.

Physical Persistence

Prima facie, physical persistence or bodily persistence seems a satisfactory criterion for the persistence of a person. A little like the existence of my grandfather clock ticking and chiming away in my office. It started life way before I was born and no doubt has gone through several owners until I bought it, rather worse for wear, in a church jumble sale and had it repaired. The clock has a spatio-temporal physical continuity and theoretically its history can be traced back to its construction in the 19th century. Unlike its present owner at this moment in time, it has had new parts added as old ones failed. In fact, it is not uncommon for old objects not to be made of their original parts as replacement parts over the years comprise the entire object. Derek Parfit (1987) provides us with the example of an old wooden ship whose planking and decking is changed over the years as they rot to the point where the ship does not consist of the same material it started off with despite its appearance being identical.[5]

Immanent Causation

People are somewhat similar to the ship, as the cells and even smaller molecules or even smaller still, the particles that define their bodies change several times in their lifetime. However, unlike the ship, bodily micro-changes are intrinsic to the organism; a self-repairing or growth process. There is a theoretically trackable physical continuity through an interconnected causal nexus: "A body persists in virtue of the fact that the atoms that are caught up in a life-preserving (causal) relation at one time pass on that life-preserving causal relation to successive swarms of atoms." [Corcoran, 2006, 72] The term given to such a self-sustaining causal relationship, first coined by Roderick Chisholm (1964), is *immanent causation*[6] and physical persistence of the person can be defined in these terms. An alternative expression being *intrinsic causation* as opposed to *extrinsic causation*; for example, death can primarily result from the intrinsic cause of heart failure or primarily the extrinsic cause of a car accident (there is some interrelationship). Thus, perhaps spatio-temporal continuity can be the defining factor in the continuity of a person. However, with some of the imagined scenarios to follow; for example, whole brain emulation, teleportation, and brain transplants change is immediate and breaks such overlapping continuity. This is particularly relevant to the resurrection scenario I intend to present.

Transplants and Prosthetics

Amputated body parts are replaced with non-biological prosthetics and biological parts from donors are transplanted. Moreover, it is not inconceivable that the entire body could be replaced in the future as medical science progresses. It would be cold to claim that an individual was less of a person because of a prosthetic limb or they are 97% Jones and 3% Lowe because of a heart transplant. If you were entirely replaced with non-biological parts, but retained your mind in totality would you be another person?[7] A non-biological total body replacement does stretch the concept of a person somewhat given that the paradigm of a person is a human organism with noetic abilities. Nevertheless, I do not think that physicality and physical continuity can be totally dismissed from the concept of a person and their continuity and I intend to argue this case further on.

If a person is sufficiently defined as their material bodies, then the resurrection of that person seems impossible unless God gathers the very particles that made up their body, but which particles, given that they have changed several times in the person's earthly life-span.

Perhaps diachronic person-identity acquires persistence because of the materialistic character of the human animal; body/mind identity theory would suggest that person-continuity is no more than body-continuity. I do not subscribe to this approach, rather I advocate a mental/material union and I intend to argue for a similar approach to diachronic person-identity; a conjunction of the psychological and physical.

Psychological Persistence

An immediate reaction to the resurrection scenario is that Smith has enjoyed psychological continuity in the resurrection process; she can remember Old-Smith's experiences before contracting Alzheimer's disease. There is also an appearance continuity as resurrected relatives recognize Smith. However, the material that constituted Old-Smith is not the same material that constitutes the new resurrected Smith; Old-Smith has been cremated and New-Smith is made of different stuff. This perspective takes a position that diachronic person-identity is fundamentally psychological persistence or continuity; physical or bodily persistence being irrelevant. Eric Olson (1999) expresses such a position more formally in terms of a psychological relation over time: "Necessarily, for any time t and t^*, and for any x that is a person at t and any y that exists at t^*, $x = y$ if and only if x, y, t, and t^* stand in some psychological relation R" [Olson, 1999, 22] and "Necessarily, for any x that is a person at t and any y that exists at another time t^*, $x = y$ only if x is at t psychological continuous with y as she is at t^*" [*Ibid*, 41]

Memory

The psychological function of memory has been posited as the critical criterion for the persistence of identity and is primarily associated with John Locke [1632–1704] (1974):

> …yet it is plain, consciousness, as far as ever it can be extended, should it be to ages past, unites existences and actions very remote in time into the same person, as well as it does the existence and actions of the immediate preceding moment, so that whatever has the consciousness of present and past actions is the same person to whom they both belong. [Locke[1632–1704], 1974, II, Ch.XXVII, 16]

The notion of memory being the essential persistence criterion is no different from the Young-Smith/Old-Smith context; that is, when she was young fit and well compared to Old-Smith in poor mental health. If, prior to Alzheimer's disease taking serious hold of Smith, she looked at photographs of Young-Smith, she would claim "that's me…" Given that the photograph was completely different from Old-Smith's present state, then recognition would be based upon memory. Once Alzheimer's took a serious hold of Smith and she lost much of her memory she would probably not recognize herself in the photograph; perhaps she ceased to be a person at this point. If so, it could be argued that the notion of diachronic person-identity is little more than the contents of a functioning memory; the memory criterion.

Embryos and Vegetables

Applying the memory criterion, the time-line of a particular human's brain experience from embryo through normal brain function to unfortunate irreversible brain damage would imply that *they* did not exist at the beginning and latter part of their lives. On this view, personhood, defined in terms of memory, suggests that embryos and human 'vegetables' (irreversibly non-cognitive) are not persons and fall outside of the persistence question; likewise, for dead persons.[8] Thus, Smith was never an embryo or an Alzheimer's patient or a dead person, she existed as a non-person between the first two states. This seems an extreme view of persistence and is more of an indicator of what it is for Smith to be a person rather than what it is for her to persist in existence.[9]

The Memory Criterion and Transitivity

The memory criterion is also questionable on other grounds. Consider the following:

Let…

P_y be a person in their youth, and
P_m be the same person in middle age, and
P_o be the same person in old age
E be an experience that occurred to P_y

If P_m remembered E, but P_o did not then, given the memory criterion, P_m is P_y, but P_o is not P_y. This challenge was proposed by Thomas Reid (1785)[10] and is a serious challenge to memory as a necessary and sufficient criterion for person-identity; this is because identity is transitive (see Identity above). Application of the memory criterion lacks such transitivity as shown from the above example; that is, from: $\forall(x)\forall(y)\forall(z)[((x = y) \,\&\, (y = z)) \rightarrow (x = z)]$).

Substituting…

$P_m(x) = P_y(y)$ and $P_y(y) \neq P_o(z)$, therefore $P_m(x) \neq P_o(z)$.

Overlapping Memory

The above defeater could be challenged by arguing that an indirect memory or overlapping criterion applies instead of a direct one: P_o is P_y because P_o remembers the experiences of P_m at a time when P_m remembered P_y experiencing E. Derek Parfit (1987) describes direct remembering as *psychological connectedness* and the overlapping form *psychological continuity*, the latter forming a relational memory chain of psychological connectedness over time.[11]

Genuine Memory

The memory criterion may suffer another telling criticism due to the memory of an actual experience being exclusive to the person who is having them and could *seem* to them to be genuine rather than being *actually* genuine; a form of personal belief rather than personal knowledge. From the above:

1. *Premise*: P_m remembers P_y experiencing E
2. *From 1*: P_m believes he is P_y
3. *From 1, 2 & the memory criterion of diachronic person-identity*: P_m is P_y

If the proposition *P_m is P_y* is true, then P_m's belief is true and justified (given a Gettier accommodation[12]), that is P_m knows he is P_y. However, what makes P_m's belief true and therefore genuine? Not simply remembering E, as P_m may only seem to remember E. What makes P_m's memory genuine is that P_m is P_y. We would need to know that P_m is P_y prior to applying the memory criterion to determine if the proposition *P_m is P_y* is true. This, of course, is begging the question.[13]

Quasi-Memory

Perhaps *seeming* to remember E is enough to assert *someone* is remembering E if that memory is causally connected, in the normal way (causally connected rather than coincidence for example), to the remembering of E; this avoids presupposing identity. The term applied to this memory description is *quasi-memory*.[14] As a possible counter, imagine a case of part-brain emulation, where the memory of one person, say Paul is incorporated into the memory of another, say Jane. Jane's first-person experience of the memory provides her with the false impression that it is an experience she had. *Prima facie*, this seems to repudiate any form of memory criterion. However, Jane's *seeming* to remember the experience is modifiable by her knowledge that it was caused in the normal way by a past experience that Paul had and the relevant continuity of identity is that of Paul.[15]

Sleeping

A telling criticism of the memory criterion is the inability to remember experiences of dreamless sleep; for example, the number of body turns in the night. The memory criterion of person-identity would suggest that the wakeful person did not exist during these periods, which is intuitively absurd.

The Psychological Connectivity Criterion

One view of psychological continuity that incorporates the memory criterion is that of psychological connectivity. Taking the scenario above, P_o (future) is connected with P_m (now), and P_y (then) because of the psychological state that P_m is in now based upon a nexus of mental events from then to now and eventually up to the future point of the existence of P_o. Memory being one of several psychological factors in such connectivity; the experiences that P_y underwent and P_m undergoes causes the memory of it in P_o together with a host of other psychological states. It is a causal connectivity; a nexus of continuous psychological causal connections where even sleep is accounted for.[16]

Fission

Reflecting on the Jane and Paul example above, a common imaginative scenario of the multiple occupancy argument involves surgical separation of brain parts into empty skulls, in particular, the cerebrum. In my overall argument, I have employed a whole brain emulation scenario and is the divine methodology I suggest is employed in the human resurrection possibility-model (see Chapter 7). Sydney Shoemaker (1984) also employed an emulation argument.[17] The imaginary notion of multiple persons resulting from an original person whether by brain surgery, brain emulation, or teleportation is termed *fission.*[18]

Brain Emulation

To emulate the brain of a person P1 in another (say P2) requires an alteration of P2's brain to precisely match the brain of P1. If this could be done in one swift action then I suspect, at the precise moment of total emulation, P2 would be a *psychological* doppelganger of P1; that is, their mental states would be identical at that moment. Under the terms of the psychological continuity criterion, P2 would terminate at the point of emulation and a new person would begin existence who is identical to P1; that is, in psychological persistence terms. In short, there are two P1's at this point; P1 and New-P1. There is separate physical continuity of P1 and P2, P1's body continues on as before and New-P1 continues on with the body of P2. Thus, P1 and New-P1 would always be different in terms of appearance and material making up their bodies.

As soon as P1 and New-P1 begin to experience new inputs, their brain states will diverge and they will no longer be psychologically identical. Divergence increases as time proceeds, but both share a psychological history at the point of emulation; that of P1.[19] The whole process is termed *branching* and despite later psychological divergence, New-P1's psychological history tracks back to that of P1's; that is, the psychological continuity line can, theoretically, be traced back to P1 at t_1 via t_2.

At the point of emulation, it could be argued that this is just a branch off of P2's psychological time-line rather than P2 ceasing to exist. This seems difficult to reconcile with psychological continuity, as P2's psychological history at the point of emulation tracks back along the P1 line, not their own. It could also be argued that New-P1's psychological future is based upon a psychological fresh start at the point of emulation rather than branching back into P1's psychological history. This too seems to eliminate the obvious psychological history inherited from P1. Consider Table 6.1:

Table 6.1. Psychological Continuity and Brain Emulation

Time:-	*t_1(pre-emulation)*	*t_2 (emulation)*	*t_3 (post-emulation)*
Person P1	→P1→	→P1→	→P1→
Person P2	→P2→	↓	
Person P3		New-P1→	→P3→

This table demonstrates an imaginary whole brain emulation scenario for diachronic person-identity for three persons over the time-line t_1 to t_2 to t_3 and beyond.

Given the psychological continuity criterion, P2 terminates at t_2 as at that time-stage New-P1 has a psychological continuity time-line back to P1 at t_1. However, at t_3 New-P1 psychologically diverges from P1 and, because of this divergence, it may be better to refer to New-P1 as P3 at t_3; that is, a different person. However, given the psychological continuity criterion P3 is actually P1, but this seems incongruous because P3 is psychologically different to P1 at t_3 even though P3's psychological time-line can be traced back to P1 at t_1 via t_2. This imaginative scenario is quite a challenge for the psychological criterion of diachronic person-identity.

Dualism

There is a body≠mind distinction suggested by Table 6.1, raising the contentious issue of dualism. If continuity of identity is just psychological continuity without biological continuity then a person is their mind irrespective of what body, biological or synthetic, they may occupy; this has the hallmark of

Cartesian dualism which is generally rejected. In so rejecting such dualism, body takes a more central role in the analysis. Take Smith; Young-Smith is a healthy thinking biological organism; the paradigm of our conception of a human person. Although her mind is not identical to her brain, it emerges from it; in this way mind and body are irrevocably bound: property dualism. Although pinning down diachronic person-identity is enigmatic, the one thing that seems clear is that person-identity comprises a conjunction of mind and body and diachronic person-identity requires reference to both.

Physical and Psychological Persistence

From a materialistic view of mind, where mental states supervene on brain states, the whole branching process shown in Table 6.1 can be viewed as a physical change resulting from the whole brain emulation procedure.

The brain part of P2's body undergoes a radical change with the emulation procedure at t_2 matching the physical brain structure of P1 at that time. Such a material change at t_2 results in a supervening mental state in P2 that matches P1at t_2; this includes memory, beliefs, and intentions and other mental states. However, apart from the brain, that is the only similarity with P1's body; overall they are very different.

Physically P1 never fully matches P2 and only part-matches at t_2 and even then, it is only a structural match not a material match; brain matter in both are still very different. Mental states only match at one point in time; that is t_2. It seems that P1 and P2 remain different persons with their mental states matching at one point in time due to a physical intervention.[20]

There is a t_3 back to t_1 psychological continuity with a branch at t_2 and a non-branching physical continuity from t_3 back to t_1. Perhaps our dissonance to the idea of different persons post-emulation can be resolved by the suggestion that a new person P_3 emerges at t_2, but this is an unnecessary complication as the physical and psychological continuities remain the same. Thus, if mental states cannot be entirely separated from physical states, then perhaps they should be viewed as a union making diachronic person-identity, a conjunction of the psychological and physical.

Personhood

Analysis of the necessary and sufficient conditions of persistence or diachronic person-identity leads to a refinement of what is meant by being a person; personhood. As I argue above, the psychological continuity criterion may be a necessary criterion for personhood, but it is not sufficient. The type of psychological factors included have a bearing on the concept. For example, self-awareness (first-person perspective), memory, rationality, language, moral understanding, complexity in knowledge, beliefs and skills, intentions, desires, strategy, and many other properties are part of the psychological mosaic that would be included in the psychological continuity nexus of a person; a complex mental/noetic system.

The First-Person Perspective

I should like to say a little more about what I have termed the first-person perspective as it is an important notion in any resurrection argument.

There are two characteristics of this perspective relevant to person-identity; a mental state of self-realization and self-reference.[21] The former being a quintessential mental state for a resurrection hypothesis; it represents that inner state of mental affairs that warrants the exclamation "it's me!" The use of the expression "it's me!" to oneself is an unusual *exclamation* and would only normally be uttered in those circumstances of surprise, such as a resurrection or a teleportation scenario (see below). It is a conscious state of self-realization or self-awareness only available from the first-person perspective.[22] It is not intended as an epistemic statement such as "I am me" which is a tautology or "I am Napoleon" which is false, although the referent 'I' is true.[23] Such statements are of the latter kind; that is, self-referring; believing one is something or someone from a first-person perspective. Self-referring from the first-person perspective is somewhat different from other-referring from the third-person perspective; both are relevant to person-identity as together they can truth-track by consensus, but not necessarily so. For example, if I believe I am Napoleon, a friend is quite liable to put me straight on that unless he is also deluded.

In terms of person-identity, first-person self-referring emphasizes the psychological, such as personal memory, whereas third-person other-referring emphasizes the physical such as appearance. Consider the diachronic person-identity of Mr. P. His first-person perspective of his own identity over time primarily engages with his past psychological states such as his memories; memories of himself as a younger man. The third-person perspective of Mr. P by Mrs. P is primarily physical such as his immediate appearance, but also her memories of life with Mr. P in the past; but they are *her* memories of Mr. P not *his* memories. Moreover, those memories of Mr. P are primarily corporeal; she will see, in her mind's eye, Mr. P in various past situations; for example, their wedding. The first- and third-person perspectives in diachronic person-identity are both of significance and demonstrate the importance of the psychological and physical criteria in combination.[24]

Phase Sortal

There is a distinction to be made between being a person and being a human. For a materialist, the latter is an animal that has or had biological functions including brain functions; even dead human bodies are still treated with reverence. People are humans and humans are animals, but we can and do make a person/human distinction. Thus, a person need not be a human and a human need not be a person; aliens, androids, and angels could be seen as persons but not human. Likewise, zombies, fetuses, cadavers, and human 'vegetables' can be viewed as humans but not persons. Thus, it seems that personhood is not a substance; if *X* is a table then *X* is not a substance by virtue of being a table, because *X* could have been a stack of wood and could become a pile of ash. *X* is a substance by virtue of being wood; planks of *wood*, a *wooden* table, and *wood* ash. If personhood is a substance, then if adult-P is a person, embryo-P is a person, and human-vegetable-P is a person, and this is generally denied.[25] *Biological organism* would be a better candidate for the substance of which *human personhood* is a sort. Thus, the predicate *person* is better described as a *phase sortal* [26] of a being, a being that could be human, but not necessarily so; a being that can exist at one moment as a person and at another moment not so. Thus, to claim Jones is a person is to claim that the human organism called Jones is of the phase-sort 'person'; that is, a conjunction of psychological qualities such as rationality and a first-person perspective. Notwithstanding, if personhood is a type of thing; for example, a human or an alien, physical existence with immanent continuity is a requirement. Hence, unless Cartesian dualism is true, there does seem to be a physicality requirement of being a person.[27] Consequently, diachronic person-identity cannot be separated from physical-continuity.

It could be argued that the biological organism substance is of the person sort at one time and not at another, psychological continuity not existing at these moments; that is, embryo-organism in the birth time-stage, then person-organism in the child/adult time-stage, then possibly 'vegetable'-organism in the old age time-stage then dead-organism in the dead time-stage. With this phase-sortal view, a person only exists in the child/adult time-stage and it is in that period that psychological continuity counts. The immanent physical continuity of the biological organism of the human sort exists through all four phases, but it is only when in conjunction with the psychological continuity criterion that personhood exists.[28] A similar, but not identical scenario would apply to aliens, androids, and angels. This conjunction of the psychological and physical is essential to provide personhood with an ontology.

Multiple Occupancy

Imagine a whole brain emulation process whereby two people resulted from one original by a whole brain emulation process; three psychological doppelgangers. Given the psychological continuity criterion there would be three organisms as one person. From Table 6.1 above, P1 need not branch to be just

New-P1, but several; that is P1's brain can be emulated in several bodies and not just biological bodies. However, psychological continuity does not entail biological continuity or even the existence of a biological organism; a third being could be an android.

First and Third-Person Perspectives

Consider the above P1/P2 emulation scenario shown in Table 6.1, but where an android is also caused to match, though synthetically, P1's brain at t_2. What would the individuals think of themselves (first-person perspective) and what would they think of the others, and what would other people think of them (third-person perspective)?

All the individuals at t_2, the emulation point, think they are P1 and if this first-person perspective were the only criterion for person-identity, then there would be one person in three separate bodies individually thinking they are the same person. This is known as *the too many thinkers' problem* and such a position, in terms of person-identity, is intuitively unacceptable.[29] A third-person perspective of this strange scenario and even P2 and the android looking at their reflection in a mirror, would clarify matters fairly quickly. Thus, person-identity seems to be a combination of first- and third-person perspectives; that is, a conjunction of the psychological character of the individual as understood by each of them and the physical character of the individual as seen by others, or by them in the mirror.

Person-Identity Revisited

As I argue above, first- and third-person perspectives of person-identity are somewhat different, with the first-person perspective placing emphasis on the psychological and the third-person perspective placing emphasis on the physical. Given these two perspectives and given the emphasis that each place on the psychological and physical properties of personhood respectively, then constructing a definition of person-identity must incorporate a conjunction of the psychological and physical. Thus, taking into account the argument so far, I define person-identity as follows: *A unique being with a complex mental/noetic system and physical body.*

Note that such a definition would include androids, aliens, and angels (provided angels are corporeal rather than phantasmal[30]) as well as humans, but exclude zombies, fetuses, cadavers, and human 'vegetables.'

Congruity in Continuity

I have argued that both psychological and physical (psycho-physical) properties are essential to the concept of person-identity and consequently essential to diachronic person-identity. Given that I have defined person-identity as *a unique being with a complex mental/noetic system and physical body*, then I would suggest the best way of expressing diachronic person-identity in terms of congruity in continuity is as follows:

> Necessarily, for any x that is a person at t; that is, a unique being with a complex mental/noetic system and physical body, and any y that exists at another time t*, x = y if and only if x is at t immanently psycho-physically congruous in continuity with y as she is at t*.[31]

Congruity in continuity means that the intrinsic psycho-physical properties of a person at different time-stages remain causally interconnected. Thus, for diachronic person-identity, neither the intrinsic psychological properties nor the intrinsic physical properties within the time-line of the two stages can become causally interconnected with a different person between these stages, nor cease altogether. Identity through congruity in continuity is a form of *best candidate theory;* that is, out of all the possible candidates for diachronic person-identity, the best candidate, in my view, is the one with the best match in terms of intrinsic psycho-physical continuity.

The Closer Continuer Theory

The above best candidate theory by congruity in continuity echoes Robert Nozick's (1984) *closer continuer theory*. Nozick argues that the best candidates for diachronic person-identity out of a set of possible alternatives are the ones with the highest degree of spatio-temporal and qualitative continuity; that is, causally related over time. For Nozick, lack of continuity of any candidate or equal closeness of multiple candidates entails cessation of continuity; therefore, there is no successor to the original; the person ceases to exist from that moment. Nozick gives weighting to longevity in his *closest* analysis; the longer-lived the closeness of continuity the better the candidate for diachronic identity.[32]

The Ship of Theseus

With reference to this closer continuer view (CCV) of identity continuity, consider the Ship of Theseus paradox.[33] In a possible world (PW_1) the old planks of the ship in dry dock are removed plank-by-plank and are reassembled plank-by-plank with new ones to form the ship again. However, the old planks are collected as they come off the original ship and are reassemble plank-by-plank *as they are removed* to form a second similar ship. Because of closer continuity to the original, CCV would have the new-plank-ship (O_1)

identical to the original over the old-plank-ship (O_2), even though they appeared to be the same ship. However, consider another possible world (PW_2) where the ship is completely dismantled and ceased to exist, *but later* reconstructed with the old planks to form the ship (O_3) so that it was the same as O_2 in PW_1. Comparing O_2 in PW_1 with O_3 in PW_2, there is no *intrinsic* difference yet, because of closer continuity to the original, CCV would select O_2 over O_3 as identical to O_1 despite no intrinsic difference.[34] Thus, if $O_2 = O_1$ and $O_3 \neq O_1$, then $O_2 \neq O_3$. Now consider the identity of indiscernibles referred to above, formally expressed in terms of the two old ships thus: $\forall F(FO_2 \leftrightarrow FO_3) \rightarrow O_2 = O_3$ (where F is a non-relational property). However, from the above example, CCV states that $O_2 \neq O_3$; there is a contradiction.

Harold Noonan (2003) argues that $O_2 \neq O_3$ is just a mere Cambridge change; that is a change in truth-value without an intrinsic change in reality.[35] For example, my car is the second fastest car in the car park; the fastest car is scrapped so my car becomes the fastest without any intrinsic change to it; a Cambridge change.[36] Thus, nothing has really happened to O_2 and O_3 in terms of their intrinsic properties, $O_2 \neq O_3$ is due to their relation to O_1; a mere Cambridge change with identity of indisernables reaffirmed and CCV questionable.[37]

Noonan's critique of CCV does not apply to the congruity in continuity view; the psycho-physical properties are intrinsic to the person. Thus, the congruity in continuity view as a best candidate theory holds.

The Only x and y Principle

The above reference to intrinsic properties relates to the indiscernibility of identicals; that is, $x = y \rightarrow \forall F(Fx \leftrightarrow Fy)$ where, in particular, F are non-relational properties. In terms of diachronic person-identity, this has been expressed as *the only x and y principle.* The principle states that person x at time t is the same as person y at t* by virtue of the relation between their intrinsic properties rather than extrinsic properties; hence, my use of *intrinsic* in the definition of diachronic person-identity. For example, my car is the same as my wife's car because the intrinsic properties of my car are identical to the intrinsic properties of my wife's car; that is, we share one car. Such identity is not the result of an extrinsic property such as the fact that my wife and I share a car or the fact that *we* are a one-car family.[38]

Applying Diachronic Person-Identity

I shall now apply the various criteria in the above definition of diachronic person-identity to four cases of imaginary fission:

1. Whole brain emulation: explained above.
2. Whole brain transplant: the brain donor dies and a recipient of the donor's brain lives on thinking they are the donor.[39]
3. Split-brain: as for 2, but the brain is divided and there are two recipients both thinking they are the donor.[40]
4. Teleportation: a person (the original) is scanned and recorded in incredibly fine detail and the data is transmitted by radio waves to a receiver that perfectly regenerates the original from new matter in every detail (the replica). The process terminates the original, but the replica is physically and psychologically (given the materialistic concept of mind) identical to the original except for the matter constituting her.[41]

Consider the following:

Table 6.2. Imaginary Fission Comparisons

Criteria↓ *Fission→*	*Emulation*	*Transplant*	*Split Brain*	*Teleportation*
Unique being	x	√	x	√
Mental/noetic system	√	√	√	√
Physical body	√	√	√	√
Psycho-physical congruity	x	*Partial*	*Partial*	√
Immanent causation	x	*Partial*	*Partial*	x

This table compares imaginary fission scenarios against the proposed criteria for diachronic person-identity. The table indicates where each scenario meets, fails to meet, or partially meets each criterion.

In the case of the transplant and split-brain scenarios, there is *full* immanent psychological causal continuity, but only *partial* physical continuity as the brain is transplanted. The table indicates this with the word *Partial*. Full-brain transplant may hold for diachronic person-identity, but the split-brain case does not because of the uniqueness requirement.

Teleportation is difficult; there no immanent psychological and physical causation; it is extrinsically caused. Indeed, this immanent causation problem with teleportation is exemplified when the original does not die for "It is logically possible that a body just like mine constitutes someone else who has a first-person perspective that is qualitatively indistinguishable from mine, but that first-person perspective would not be mine. Nor would that person be me." [Baker, 2000, 133]

Of the four imaginary examples, only full-brain transplant comes close to fulfilling the diachronic person-identity criteria, but only just. It is understandable why there will be intuitive disagreement on these two imaginary scenarios.

CONCLUSION

In this chapter, I presented a view on what counts as person-identity and diachronic person identity; both are related and an understanding of both are essential to any resurrection hypothesis. The person-identity view reflects a particular understanding of mind; that is, a mental-material union. Indeed, I have taken the view that both the psychological and physical perspectives of person-identity are essential to the concept, defining person-identity as: *A unique being with a complex mental/noetic system and physical body.*

Such a definition would include androids, aliens, and angels as well as humans, but would exclude zombies, embryos, cadavers, and human 'vegetables.'

My view on diachronic person-identity echoes that of person-identity per se, being a conjunction of the psychological and physical criteria and being expressed as:

Necessarily, for any x that is a person at t; that is, a unique being with a complex mental/noetic system and physical body, and any y that exists at another time t, x = y if and only if x is at t immanently psycho-physically congruous in continuity with y as she is at t*.*

As can be seen from the above, the relationship between the psychological and physical is one of being congruous in continuity. By this I mean, in the time-line of a person's identity between a later time-stage and an earlier time-stage; that is, looking at their history, neither the intrinsic psychological nor physical properties branch off to alternative beings or cease altogether in that time-line.

Immanent causation is also an essential criterion in diachronic person-identity; that is, person-identity between time-stages results from a causal chain intrinsic to the person.

Armed with these two expressions of person-identity and diachronic person-identity, I intend to consider the possibility of human resurrection in the next chapter.

NOTES

1. Note I employ the term *person-identity* rather than *personal identity* as the latter implies identity applied to oneself; the former has a greater scope.

2. A non-relational property is intrinsic in so far as the property is not acquired because of its relation to something else. For example, *redness* is an intrinsic non-relational property, whereas *large* is a comparative relational property. This relates to the only x and y principle introduced further on.

3. Leibniz W. G. [1696] (1973); p. 133ff.

4. Transitivity in identity is based upon intrinsic properties of the component things; that is non-relational. Thus, the notion of *morning* and *evening* is not relevant to transitivity in identity.

5. Parfit D. (1987); p. 203f; this transformation is associated with the Ship of Theseus paradox.

6. Chisholm R. (1964); p. 7 and Zimmerman D. (1997); pp. 433–471.

7. Baker L. R. (2007); p. 167.

8. Mackie D. (1999).

9. *Ibid;* pp. 224–228.

10. Thomas R. (1785)—*Essays on the Intellectual Powers of Man*—Essay VI; Ch. 5. Reid provides the example of child scrumper who becomes a brave military officer and eventually a General. The officer remembers the scrumping incident and the general remembers the brave officer, but not the scrumping. If memory is the persistence criterion, is the child, the officer, and the General the same person?.

11. *Ibid*; p. 206f.

12. I refer to Edmund Gettier's acclaimed 1963 paper *Is Justified True Belief Knowledge?*

13. The presupposition of person-identity to define person-identity is associated with Bishop J. Butler (1736)—*The Analogy of Religion*

14. Shoemaker S. (1984); Ch. 2 and Parfit D. (1987); p. 220.

15. Parfit D. (1987); pp. 220–222.

16. Olson E. (2007); pp. 17–24.

17. Shoemaker S. (1984)—*Personal Identity: A Materialist's Account* in Shoemaker and Swinburne, *Personal Identity*, Oxford: Blackwell.

18. *Ibid*; pp. 117–119 and Noonan H. W. (2003); pp. 139–142.

19. Compare Parfit D. (1987); pp. 253–261.

20. This fulfills the uniqueness requirement of person identity; i.e., a person can only possess singular existence at any one time, i.e., there can only be one 'you': Reflexivity above and Olson E. (1999); p. 49.

21. One can, of course, have a first-person perspective on things and events other than oneself, but self-reference is specific to person-identity.

22. Baker L. R. (2000); Ch. 3.

23. *Ibid*; pp. 70–73.

24. Evans G. (2000); Ch. 7 for further discussion on self-identification.

25. *Ibid;* p. 29.

26. A phase-sortal is an identity descriptor of a thing that applies at certain times during its existence and not at others. A sortal descriptor states what something is and provides both a classifiable tally and a continuity requirement. For example, 'a footballer' is a type of sportsperson, is one player, and continues to be a footballer provided she plays football regularly. 'A footballer' is also a phase-sortal in so far

as she did not play as an infant and is very unlikely to play in old age: Wiggins D. (1980); pp. 24–27.

27. As a materialist I believe even angels can be viewed as material beings; just not in this world.

28. It should be noted that a conjunction is false if one conjunct is false.

29. Noonan H. W. (2003); pp. 209–213.

30. Given the materialistic argument for Heaven, then the physicality of angels is feasible.

31. Compare Olson E. (1999); p. 41; note that this definition of person-continuity is in accord with the principle of the Indiscernibility of Identicals defined by: $x = y \rightarrow \forall F(Fx \leftrightarrow Fy)$.

32. Nozick R. (1984); p. 43ff.

33. Theseus was a mythical king of ancient Athens who killed the Minotaur in Crete; the ship was the vessel he returned in from Crete.

34. This is known as the Ship of Theseus paradox: Wiggins D. (1980); pp. 92–94.

35. Geach P. T. (1969)—*God and the Soul*—London: Routledge.

36. Slors M. (2004); pp. 11–15.

37. Noonan H. W. (2003); p. 138f.

38. *Ibid*; pp. 129–131.

39. Olson E. (1997); pp. 44–46.

40. *Ibid*; pp. 46–52.

41. Parfit D. (1987); pp. 199–211.

Chapter 7

The Field

The Possibility of Resurrection

In this chapter, I will muster the conclusions of the previous chapters in support of an argument for the resurrection hypothesis by way of a theoretical model to establish the *possibility* of human resurrection.

Resurrection as a life after death scenario could be a natural process; that is, a scientific enterprise and I present some possible scenarios along these lines. However, it can also be one requiring divine intercession and this is the main tenet of my resurrection hypothesis. The natural process following death is decay and there is little if any evidence of an immortal soul in the form of a non-corporeal spirit that leaves the body on death to continue a spiritual life in Heaven.[1] Thus, I propose a materialistic resurrection hypothesis that requires the will of God to instantiate it.

Given the necessity of divine intercession in the resurrection process, there appears to be two perspectives involved: intention and method. The former being divine intention; that is, assuming God exists, for resurrection to be an actual state of affairs, God would need the desire and then the wherewithal to instantiate it. In Chapter 1, I argue that God cannot do the logically impossible and human resurrection is not logically impossible; however, logical impossibility is but one of a nest of impossibilities. Conceptual, metaphysical, and nomological impossibilities also loom over the resurrection hypothesis. This raises the issue of method; supposing God does have the desire to resurrect humans, how can such a desire be fulfilled; that is, what method would be employed? In this chapter, I intend to address these issues and argue that God, if He exists, would have both the desire and the wherewithal to resurrect humans after their death. Moreover, I intend to guide the resurrection hypothesis through the maze of the aforementioned impossibilities and finally present a model for a resurrection process; a model with genuine possibility.

The resurrection model posited for genuine possibility is complex and may be difficult to conceive of in practical terms. However, it is simply a model to demonstrate that human resurrection is a genuine possibility rather than an impossible whim of the theist. It is not a claim that the detailed content of the model is what *actually happens*, rather it is a claim of *what could happen* without denying that other models may also be genuinely possible.

Divine Desire

In Chapter 1, I argue that God necessarily has the attribute of omni-benevolence and I defined this attribute as follows:

> *The necessary maximization and amelioration of all states of affairs that result in moral truths and moral falsities respectively with the ultimate aim of realizing the greatest possible good.*

Michael Hauskeller (2016) considers death as making life meaningless. Indeed, death is a natural evil in so far as it deprives persons of their life, future pleasures, achievements, aspirations, and hopes and renders all we do as meaningless; that is, without purpose.

> It is easy to feel that if one day, however far in the future that may be, everything is over for us and, worse still, nothing will remain of what we have done and achieved in this life, then our hopes and aspirations are actually quite hopeless. [Hauskeller, 2016, 2]

I think this is a little strong; however, there is much to lose through death, particularly if it is premature. By contrast, Bernard Williams (1973) argues that immortality can also be an evil through inevitable boredom; thus, "…an endless life would be a meaningless one; and that we would have no reason for living eternally a human life." [Williams, 1973, 89]

However, boredom need not be inevitable in the immortal context, even in this world repetition is not always boring; for example, inspiring music can still be inspiring no matter how many times it is listened to. Beautiful scenery can be taken for granted if it is continuously observed, but it can maintain its aesthetic impact if there are observational breaks. Purposeful activity need not have a finite time-line and loving relationships are always inspirational; I remain unpersuaded by Williams' paper.[2]

In Chapter 4, I argue that, if God exists, He may permit moral evil and natural evil like death for the greater good; for example, by withdrawing His presence for the purposes of human free-will. However, if such natural evil, such as death can be ameliorated without loss of the greater good attained by its existence, then an omni-benevolent being would necessarily desire this. If resurrection is a possibility, then it provides for this opportunity; that is, death can be reversed without a direct presence in this world and the inevitable loss of free-will. In the same vein, resurrection would also go some way to ameliorating the moral evil of murder and other such abusive crime without the need for the hand of God to prevent the evil acts.

Divine Power

In addition to omni-benevolence, I also argue in Chapter 1 that God has the attributes of omnipotence, omniscience, and divine freedom defined as:

> Omnipotence: *The capability to do anything that is possible to do better than any other being*
> Omniscience: *Elected knowledge of every possible knowable thing*
> Divine Freedom: *The choices of an omnipotent being to do anything that is possible to do are totally unrestricted*

With such attributes and the desire to ameliorate the natural evil of death through resurrection, God has the wherewithal and the will to instantiate the state of affairs that enables human resurrection. However, note the caveat in the definition of omnipotence; that is, '*...anything that is possible to do...*' The issue of possibility now needs to be addressed.

GENUINE POSSIBILITY

The resurrection hypothesis needs more than conceivability, it needs genuine possibility and Scott Sturgeon (2000) argues that there are four criteria that are candidates for genuine possibility:

1. *The formal criterion*: possibility is ensured by logical form.
2. *The conceptual criterion*: possibility is ensured by the meaning of the concepts employed.
3. *The metaphysical criterion*: possibility is ensured by the essential nature of the concepts employed.
4. *The nomic criterion*: possibility is ensured by the laws of nature.[3]

Sturgeon's criteria are an objective approach to assessing *impossibility*, which is pertinent to a resurrection argument based upon the will of God given that I have argued that omnipotence is limited by the impossible. Consequently, I intend to assess the resurrection hypothesis by way of a model based on the conclusions of the previous chapter that may fulfill Sturgeon's criteria. I also intend to employ the proposition *Smith died and was resurrected* as a case in point for the formal and conceptual criteria.

The Formal Criterion

The formal criterion will exclude the possibility of the truth of propositions that are contradictory. For example, *Smith is dead and not dead* is a contradiction; however, although *Smith died and was resurrected* is related to *Smith is dead and not dead*, it is not a contradiction. Indeed, the proposition is closer to *Smith died and is now not dead* which also is not a contradiction. Thus, the resurrection proposition passes the formal criterion; that is, it is not an impossibility on the basis of that criterion.

The Conceptual Criterion

The conceptual criterion precludes genuine possibility if the meanings of concepts contradict each other. The concepts of death and resurrection in the proposition *Smith died and was resurrected* do not appear to be contradictory. However, the proposition *Smith survived her death* does appear to be contradictory and *prima facie*, equivalent to the resurrection proposition.

The proposition *Smith survived her death* may not be contradictory given the example of a resuscitated patient who medically died; this seems to be a case of surviving death. However, the noun *death* in this context is a technical term meaning the heart stopped or some other physiological phenomenon and the patient is brought back from the brink. This is somewhat different to common usage where *death* means *dead and gone*; that is, beyond the brink.

The proposition *Smith survived her accident* suggests Smith did not die in her accident. By contrast, in the resurrection case, Smith did die as resurrection presupposes death, whereas survival presupposes the denial of death as in the case of the accident. Thus, a more accurate rendering would be *Smith survived after her death*, which is not contradictory given the notion of life after death; hence, the resurrection proposition passes the conceptual criterion.

The Metaphysical Criterion

The metaphysical criterion precludes genuine possibility if the essential properties of the concepts employed in the proposition are incongruous. Thus, for the proposition *Smith died and was resurrected* to be metaphysically possible it must be capable of being true in some world. The relevant concepts in the proposition are:

1. Personhood (Smith is or was a person in an Earthly time-line)
2. Death (Smith died)
3. Resurrection of the person (Smith 'rose from the dead')
4. Identity (Resurrected Smith is identical to the ante-mortem Smith)

Personhood and death are uncontentious, but resurrection and identity are. This is a case of diachronic person-identity through a death time stage and pulls upon the ideas and arguments of the previous chapter. Therein, I argue that diachronic person-identity can be expressed as: *Necessarily, for any x that is a person at t; that is, a unique being with a complex mental/noetic system and physical body, and any y that exists at another time t*, x = y if and only if x is at t immanently psycho-physically congruous in continuity with y as she is at t*.*

Given this, the resurrection hypothesis should satisfy the following four conditions for resurrected Smith to be identical to the ante-mortem Smith:

The resurrected Smith must be…

1. A unique being
2. A physical being
3. In possession of a complex mental and noetic system
4. Psycho-physically congruous in continuity with the ante-mortem Smith
5. Causally (immanent causation) connected to the ante-mortem Smith

Fission

One way of engaging with the metaphysical possibility of resurrection is to reflect upon the imaginary fission examples of diachronic person-identity, particularly those close to fulfilling the above criteria.

In Table 6.2 in Chapter 6, I suggested that the only imaginary fission candidate for diachronic-person identity was the brain transplant and this was questionable in completely satisfying the fifth criterion above.

Brain Transplant

In the case of the brain transplant, psycho-physical congruity in continuity is not perfect; that is, the brain-recipient's psychological and physical time-line does not follow the same historic track; it branches. The brain-recipient's psychological history tracks back to the brain-donor's past, but only part of the brain-recipient's physicality tracks back to the brain-donor's past; that is, only the brain part. The brain-recipient's body (sans brain) tracks back to the brain-recipient's own physicality prior to the brain transplant. Thus, there is partial psycho-physical congruity in continuity in the sense of mental/noetic and brain continuity. It could be argued that this is sufficient for diachronic person-identity; some would beg to differ.

Teleportation

Teleportation is quite different, but there is still a problem with the fourth criterion; not because of branching as in the brain transplant case, but because of the immanent causation element of the criterion. The immanent causation requirement states that physical and psychological changes in a person over time, say from childhood to adulthood, must result from intrinsic causation and not from an extrinsic source.

The teleportation argument, briefly introduced in the previous chapter, involves a futuristic machine that can scan every detail of a human body (the original) in incredibly fine detail. This scan is then converted into electronic data that is transmitted by electromagnetic waves to a receiving station at some distance that converts it back into a perfect replica of the original. In the teleportation process the original is destroyed, but the replica is a perfect copy and given a materialistic conception of mind, the replica's mind emerges matching that of the original; even the first-person perspective, as the incredulous replica announces "it's me!" The matter constituting the original is lost and the matter constituting the replica is new. There is congruity in continuity and it might seem that the replica is the original except that the causal process of teleportation is extrinsic and not immanent, giving the intuitive conception that the original died and the replica is a new person; not the original.[4]

It could be argued that there is a causal link between the original and replica *mediated* by the teleporting process, but this is more a case of transient rather than immanent causation. There is also extrinsic mediation between a child maturing into an adult; that is, the child's environment must be conducive to growth. Nevertheless, the teleportation is immediate, not piecemeal, and the original is destroyed, whereas the child matures. Claim to diachronic person-identity by teleportation is a gray area mitigated by the fact that the replica is convinced she is the original and third parties, unaware of the teleportation process would believe this too; both could be false beliefs. Teleportation by particle entanglement has more of an immanent causal character than the radio transmission example above and has already occurred at the particle level.[5]

Noetic continuity in teleportation may be problematic, as it appears that the cause of the replica's memory, beliefs, and character are not genuine.

For example, remembering experience E because E *actually happened* to the replica as it did the original.[6] It did not happen to the replica, the memory has been caused by the teleportation process.

The teleportation example is even more intuitively perplexing if, because of transmission problems, the original is immediately reconstituted as a precise replica in the same teleporter using the same material that comprised the original; everything is identical, including the very particles. The identity sceptic may be tilted more towards diachronic person-identity through teleportation given this intuitive scenario. Notwithstanding, immanent causation is a serious issue for diachronic person-identity through distance teleportation.

The Emulation Problem

There are teleportation arguments that consider the situation where the original does not die; however, this would be a breach of the uniqueness criterion given that there are two persons claiming the same identity. Moreover, interesting though these arguments may be, they have little application to the resurrection hypothesis given that in resurrection one person is dead. On this basis, I should like to reconsider the whole-brain emulation scenario presented in Table 6.1 in the previous chapter, rejected there because two or more persons are claiming the same identity. However, if P1 dies at t_2, the point of emulation, then the uniqueness criterion may be satisfied. However, that still leaves a breach of the psycho-physical congruity of continuity criterion, as P3's psychological continuity branches off at t_2 tracking P1, whereas her physical continuity tracks to P2. Moreover, like the teleportation process, the same immanent causation problem remains because of the extrinsic emulation process.

Brain and Body Emulation

One possible solution to the branching problem is to introduce a reversible non-cognitive human organism into the process that is receptive to body emulation in the way the brain is in my earlier example.

Consider the table:

Table 7.1. Body and Brain Emulation

Time:-	t_1*(pre-emulation)*	t_2*(body emulation)*	t_3*(brain-emulation)*	t_4
Person P1	→P1→	→P1→	→P1→ dies	
Organism O	→O→		↓	
Person P2		O/P1^B→	→New-P1→	→P2→

This table presents a possibility-scenario for whole body and whole brain emulation over the time-line t_1 to t_2 to t_3 to t_4 and beyond. The scenario involves two persons and a non-cognitive human organism. One person dies and the organism morphs into the other person through the emulation process.

Let…

O be a reversible non-cognitive organism (a healthy, living, unconscious humanoid organism); a potential counterpart to P1
P1 be person 1
P2 be person 2
P1 die at t_3 post-brain emulation
^{B} stand for body

Upon body-emulation at t_2, O physically matches that of P1 (shown as O/P1^B in the table); however, O is still unconscious and the matter constituting the body of O/P1^B is different to that of P1. O/P1^B's brain upon brain-emulation at t_3 matches that of P1; again, the matter constituting the brain is different. At t_3, P1 dies and O's brain cognitively awakens matching the mental/noetic state of P1 (shown as New-P1 in the table). At the point of brain-emulation and the death of P1, the non-cognitive organism becomes a person going on to time-stage t_4 where her brain state changes due to new experiences (she becomes P2 as shown in the table). The historic time-line of P2 tracks back from t_4 to t_3 to t_2 to t_1 as P1 in terms of her body and mind albeit with different body and brain matter. Thus, there is congruity in continuity and uniqueness as there is one person at t_3 and t_4 due to P1's death.

There is, of course, the problem of creating a counterpart organism like O in the first place, but this is merely an intuitive scenario to seek metaphysical possibility.

As I argue in Chapter 6, person-identity also has a third-person perspective focussing upon appearance and mental/noetic states interpreted through behavior such as language. The reversible non-cognitive organism process accounts for this requirement. However, the problem of immanent causation remains.

Immanent Causation

The notion of immanent causation was introduced by Roderick Chisholm (1964):

> I shall borrow a pair of medieval terms, using them, perhaps, in a way that is slightly different from that for which they were originally intended. I shall say that when one event or state of affairs (or set of events or states of affairs) causes

> some other event or state of affairs, then we have an instance of *transient* causation. And I shall say that when an agent, as distinguished from an event, causes an event or state of affairs, then we have an instance of *immanent* causation. [Chisholm, 1964, 7]

In terms of diachronic person-identity, transient causation includes physical and psychological states and is extrinsic to the person. For example, the imaginative teleportation process involves extrinsic causation of the teleporting apparatus not the intrinsic causal efficacy of the original. By contrast, immanent causation is intrinsic to the person and includes physical and psychological states; for example, a child maturing to adulthood. There is frequently overlap between the two causal functions; for example, the causal nexus of the maturing child would also include the child's environment and may include life-saving medical intervention. In the case of psychological persistence, memory and beliefs are also mediated by extrinsic causal factors such as travel and education. Notwithstanding, there must be considerable immanent causation to claim persistence in person-identity, no less so than in the case of divinely caused human resurrection. "For God to give you life after death, then, he would have to move you, or some remnant of you that preserves your psychological states, along a continuous path from this world to the next one." [Olson, 2010, 57] Thus, if God instantiates a post-mortem person from an ante-mortem person entirely by transient causation, then identity would be lost. This is the case in the whole-body-whole-brain emulation example above and, like the teleportation example, diachronic person-identity is lost.[7]

Divine Body-Snatching

One resurrection hypothesis that *prima facie* resolves the immanent causal requirement, is proposed by Peter van Inwagen (1978). Reflecting on possibility rather than probability, he provides an imaginative body-snatching possibility, whereby God surreptitiously snatches a dying person at the point of death, replacing it with a simulacrum that the relatives believe is the corpse of the deceased which they bury or cremate. The dying person is revived in Heaven and diachronic person-identity is preserved as is immanent causation of persistence.[8]

However, this is a case of divine deception, something Descartes argues is impossible:

> …God…must possess all the perfections of which I can attain any notion… although I cannot comprehend them; and he must be liable to no defects. From this it is clear enough that he cannot be deceitful; for it is obvious by the light of nature that any fraud or deceit depends on some defect. [Descartes [1637], 2008, third meditation]

The Falling-Elevator Model

An alternative, less deceitful resurrection hypothesis that also claims to resolve the immanent causal requirement, is proposed by Dean Zimmerman (2010) and is termed the falling-elevator model.[9] The title is based upon those cartoon scenarios where an occupant of a falling elevator is saved at the last moment by leaping from the elevator before it crashes in to the basement floor. The resurrection hypothesis based upon this analogy states that a person, close to death, fissions into two; a corpse and living person with the former going to the grave and the latter leaping through space-time to Heaven. However, apart from the issue of nomological possibility, the model relies upon the soundness of the closest continuer theory; that is, the leaping person is the closest continuer to the ante-mortem person than is the corpse, consequently identity is maintained as is intrinsic causation.[10] Nevertheless, although the falling-elevator model may fulfill metaphysical possibility, leaping bodies breaking through the fabric of space-time would not pass the nomological possibility requirement, unless of course, the model relies on the *God moves in mysterious ways* argument.[11] Moreover, the closest continuer theory is questionable, as I argue in the previous chapter.

A Causal Interface

One possible solution to the immanent causation problem with the emulation example above is to undertake the emulation slowly over time while the emulator (P1 in the example) is alive and create an emulator/emulatee causal interface. In this way, emulation is enabled by this causal interface rather than the emulating technician; that is, the technician mediates the causal process, but does not break into the causal chain.

During the emulation process, the emulator is conscious and functioning normally, but the emulatee (O in the above example) is unconscious and its body dormant until the emulator dies, then it is awakened by the technician. Thus, *prima facie*, it may seem that, even prior to death, the emulator is becoming identical to the emulatee; nevertheless, while the emulator is alive the uniqueness condition would be breached. Theodore Drange (2015) argues that the *possibility* of pluralization results from new body afterlife scenarios; in the above scenario, multiple emulatees. I believe multiple emulatees would exclude identity between the emulatees;[12] however, whether the mere possibility alone would exclude person identity in a resurrection scenario is questionable.

Immanent Causation in Emulation

As stated above, this emulation scenario of Table 7.1 does not appear to be a case of immanent causation, as the changes in O caused by P1 are from one distinct thing to another. Immanent causation requires all changes to be within, or intrinsic to, a thing. For example, nuclear fission results in energy being released from matter when close to its critical mass; this is an intrinsic reaction and the released energy is a case of immanent causation, albeit with the mediation of a technician. Likewise, with the persistence of persons; the changes from childhood to adulthood are primarily within the body of the person with some outside mediation.

Aggregate Immanent Causation

In the aging process of an individual their "…body persists in virtue of the fact that the atoms that are caught up in a life-preserving (causal) relation at one time pass on that life-preserving causal relation to successive swarms of atoms."[13] [Corcoran, 2006, 72] In the P1/O scenario the atoms of P1 "are caught up in a [P1] life-preserving (causal) relation" and the atoms of O are also "caught up in an [O] life-preserving (causal) relation," but in addition their atoms are collectively caught up in a P1/O emulating (causal) relation and all these relations are passed on "to successive swarms of atoms." Because of the long-term, continuous, and synchronous causal relation between P1 and O, perhaps the P1 and O collections of atoms can be viewed as a single collection rather than two distinct ones; an aggregate. This could be likened to an immanent causal state of affairs intrinsic to this aggregate. Such an aggregate perspective over distance is supported by the notion of quantum entanglement, a spooky[14] inter-particle relationship over a vast distance:

> Once two particles become entangled, it doesn't matter where those particles are; they retain an immediate and powerful connection that can be harnessed to perform seemingly impossible tasks…it is possible to link particles together so completely that the linked objects (photons, electrons, and atoms, for instance) become, to all intents and purposes, part of the same thing. [Clegg, 2006, 1f]

In Table 7.1, P1's demise would then be viewed as the death of part of that life-preserving collection rather like an amputated limb. An analogous example of aggregate immanent causation is a pregnant mother, where the fetus and mother is generally viewed as an immanent causal aggregate. Less causally immanent, but analogous is a flock of flying starlings. The flock turns, twists, and swoops as one giant organism despite comprising hundreds of individual birds. These examples do not completely match the paradigmatic child to adult immanent causal state of affairs, but they are close.

All the other conditions of diachronic person-identity have been met; uniqueness, physicality, possession of a complex mental/noetic system, and congruity in continuity.

Given the above, it appears *prima facie*, to be reasonable to claim that P2 is identical to P1. Moreover, third parties would also recognize P2 as P1:

cognitively and physically. This brain and body emulation process paves the way for a materialist resurrection hypothesis that is metaphysically possible. However, apropos resurrection, a further problem lurks.

The Only X and Y Principle

There is a problem with the only x and y principle with this emulation process in a resurrection scenario. Three organisms are involved: the ante-mortem person, the resurrected person, and the post-mortem corpse.

The closer continuer view would support the living/resurrected identity over the living/corpse identity, but the closer continuer view is questionable.[15] Moreover, consider Harold Noonan's rendering of the 'only x and y principle':

> ...whether a later individual y is identical with an earlier individual x can depend only on facts about x and y and the relationships between them: it cannot depend upon facts about any individuals other than x and y. [Noonan, 2003, 129]

Consider once again Smith the adult, Smith the corpse, and resurrected Smith. A claim of identity between adult-Smith and resurrected-Smith breaches the only x and y principle because of corpse-Smith. A resurrection argument relies upon the fact that corpse-Smith is not identical to adult-Smith. Thus, the claim of identity does not solely depend upon facts about adult-Smith and resurrected Smith, but also facts about corpse-Smith. One possible solution to this conundrum with the resurrection possibility-model is not to claim formal identity in resurrection, but resurrection by constitution; that is, resurrected-Smith and adult-Smith are similarly constituted.

Constitutionalism

Consider the frequently used example of Michelangelo's statue of David and its constituting marble. The statue *David* and the marble *Piece* are one thing and therefore seemingly identical. However, Piece can be reshaped into something else; for example, a pillar, and if so David at time$_1$ and Piece at time$_2$ would not share non-relational properties and the identity of indisernables (Leibniz's Law; $\forall F(Fx \leftrightarrow Fy) \rightarrow x = y$) would be breached. Notwithstanding, David and Piece are existentially closely interwoven. What makes them different at one moment and seemingly identical at another is the circumstances each find themselves in. If Piece is in the circumstance of being in an art gallery, then Piece constitutes David, when Piece is in the circumstance of being a re-shaped pillar it does not.[16]

Lynne Baker (2000) introduces a concept of *primary kinds*; that is, what a thing is fundamentally. David is fundamentally a statue and what makes it a statue is its *circumstances*. In this case being exhibited in the Galleria dell'Accademia in Florence.[17] Piece is fundamentally a lump of marble which is its primary kind. Its circumstances can vary between, for example, being exhibited as art or being constructional. It is the primary kind and circumstances that, for Baker, define a thing's constitution. If not formal identity, perhaps adult-Smith and resurrected-Smith could exhibit a constitutional ontology.

For Baker, adult-Smith's *primary-kind* is personhood, that is, essentially possessing a first-person perspective.[18] Adult-Smith's *circumstance* is being a member of the living human-set and thus having the capacity for a first-person perspective. Adult-Smith is contingently constituted by a human organism; therefore, the organism need not be identical over time or across worlds.

This can be contrasted with, for example, an android. An android can essentially possess a first-person perspective because of its circumstances; that is, being a member of the functioning conscious android-set. The android is contingently constituted by a robotic mechanism which can be changed over time by a technician.

Resurrected-Smith's primary-kind is also an essential first-person perspective because of her circumstance; namely, being in the living human-set. She, like adult-Smith, is contingently constituted by a human organism. By contrast, corpse-Smith, even though being contingently constituted by a human organism is not a member of the living human-set and therefore lacks the primary-kind property of a first-person perspective. Corpse-Smith is entirely different from adult-Smith and resurrected-Smith.

In the case of David and Piece, identity is claimed when their circumstances match and Piece's primary-kind property is also possessed by David; that is, the same marble. Likewise, adult-Smith and resurrected-Smith must also share the same circumstances and primary-kind properties and, for the constitutionalist, they do. They are both part of the human-set and both have the same first-person perspective. Also, they are both contingently constituted by a human organism, but not the same organism.[19]

Resurrection through constitutionalism does not attempt to resolve the 'only x and y principle' problem (there are still relevant facts about corpse-Smith), but circumnavigates it by departing from the domain of formal identity. However, there is a pre-assumption that the first-person perspective belonging to resurrected-Smith is the same as that belonging to adult-Smith. This would be like assuming the first-person perspective of the replica is the same as that of the original in the imaginary teleportation scenario, therefore the replica is identical to the original. In both cases, the argument begs the question.[20] Nevertheless, if the whole brain emulation process in the possibility-model posited can replicate the first-person perspective of adult-Smith in resurrected-Smith, then the charge of circularity is rebuffed. If a mental state is an emergent property supervening on a brain state then, just like other mental states, such as memory and beliefs that supervene on brain states, a first-person perspective can theoretically be subject to emulation; that is, transferable.

Baker's constitutionalism does not entirely refute my notion of what constitutes a person; that is, *a unique being with a complex mental/noetic system and physical body*. It is that some of the components are contingent and one is essential within the mental/noetic system; namely, a first-person perspective.

The possibility-model posited for human resurrection does not contradict resurrection by constitutionalism, but strengthens it. Immanent causation through aggregation, whole brain and whole body emulation add to the notion of cross-world persistence, despite constitutionalism falling short of formal identity.

Resurrection

The above whole body/brain emulation process presented in Table 7.1 requires third party mediation and translating the above into a resurrection scenario, this third party is seen as God. Although it is conceivable how a P1/O immanent causal interface could be engineered in our world as both individuals can be physically interconnected, it is difficult to conceive of a cross-world connectivity: Earth/Heaven. However, such practicalities fall under nomological possibility and suffice it to say, the above resurrection scenario appears metaphysically possible; that is, there is nothing about the concepts involved that make it impossible. Consequently, the proposition *Smith died and was resurrected* is not metaphysically impossible.[21]

Nomological Possibility

The proposition *Smith died and was resurrected* has navigated around logical, conceptual, and metaphysical impossibilities, and the final hurdle is nomological impossibility.

Nomological impossibility has an empirical edge lacking in the other criteria; it is about how the world cannot be because of the laws of nature and, given the other-worldly nature of the resurrection hypothesis, it is a formidable hurdle to overcome. This is because nomological impossibility relates to the laws of nature governing this universe, not the Heavenly realm, where the laws of nature therein may be quite different. Hence, it could be argued that the laws of nature governing our universe have no application to human resurrection, some other unknown laws do and, consequently, human resurrection cannot be deemed impossible; an argument from ignorance. However, such an argument is hollow and quite unconvincing, particularly when the possibility-model being considered involves cross-world connectivity. Moreover, the existence of Heaven must be given some consideration as it is, along with our world, the setting for the possibility-model.

Human resurrection in our universe may not be a nomological impossibility; cryonics and Steinhart's digital afterlife would be examples where human resurrection is a distinct possibility. However, in this chapter I have posited a possibility-model for human resurrection in the context of Heaven and it is that model that must be considered. Moreover, if the model is to be predicated with any empirical credence, there are a set of questions that must be addressed: is an alternative world/universe an impossibility, and if not, is inter-world connectivity an impossibility? Also, given the possibility-model posited, is cross-world whole body and brain emulation also an impossibility?

Heaven

In Chapter 3, I presented the cosmological notion of multiverse theory,[22] arguing that although there is little, if any, empirical evidence for the existence of alternative universes, the theory is underpinned mathematically through quantum field and string theories. String theory or more recently, M-theory providing the possibility of multiple space-time dimensions as well as multiple universes. From the different forms of multiverse that emerge from quantum field theory, the landscape universe seems the most commensurable with the concept of Heaven. The landscape universe emerges from the mathematics of quantum tunneling; a process whereby a particle, such as an electron, can emerge on the other side of an apparently impervious barrier.[23] In short, it has been contended that quantum tunneling could result in whole universes nesting within others.[24]

One multiverse model is that of the infinite multiverse; an infinite number of universes which in turn, presupposes an infinite number of identical universes and an infinite number of near identical universes.[25] Hence, if the infinite multiverse is possible, the equivalence or near equivalence of cross-world laws is also a possibility. Combine these possibilities with the multi-dimensional space-time model of M-theory and the landscape universe concept, then a hidden close-up universe, nesting within our universe and exhibiting life sustaining laws, is not an impossibility; such a universe could be entitled Heaven.

It could be argued that multiverse theory is just a mathematical construct without scientific substance and hence lacking nomological substance. I argue in Chapter 3, that mathematical cosmology does have substance; the mathematics of Maxwell with electromagnetic waves, Einstein's theory of general relativity with gravity, and more recently gravitational waves, Higgs boson and many other verified mathematical assertions give substantial credence to mathematical assertions. Moreover, mathematics could well be an essential component of reality that is discovered not invented.[26]

Inter-World Connectivity

A further challenge to the possibility-model is the cross-world connectivity to enable human resurrection through whole body and whole brain emulation. It is difficult enough to come to terms with the technicalities of emulation in our world, but to explain cross-world causal connectivity is indeed a challenge. The advantage in the cross-world resurrection scenario is that the technician has formidable powers; such as omnipotence and omniscience. Given these powers, if it is possible to enable such connectivity and God wants to do it, then it will happen. Indeed, there is evidence of connectivity; in Chapter 5, I addressed the issue of religious experience, providing evidence of a range of different experiences to probabilistically confirm the existence of God. Our brains seem to be sensitive to this divine connectivity and given the efficacy of prayer, so too is God in the other direction. However, the resurrection possibility-model I have proposed goes beyond connectivity per se; there is an aggregate immanent causal connectivity.

Inter-World Causation

At the particle level in our universe, it is indeed possible for one group of particles to cause another group of particles to correlate with them. This is the case in superconductivity,[27] where electrons, at considerable distance from each other, correlate to form pairs that rotate around each other at the same speed, but in opposite directions and match in a head-to tail-spin.[28] This is the process of particle entanglement of quantum mechanics; that is, entangled particles exhibit a correlated state when measured at a particular moment, no matter what the distance there is between them.[29] However, entanglement is a phenomena of our universe, the possibility of inter-world entanglement is a different issue.

The above nesting landscape universe model may provide the proximity requirement for cross-world connectivity; however, there is still an inter-universe border to contend with. M-theory states that sub-atomic particles like electrons, photons, and neutrons are defined by strings: miniscule vibrating energy strands. M-theory, proposes two types of strings; loop and snippet and sub-atomic particles comprise strings of these two types. However, snippet strings are bound to their universe, whereas loop strings, such as those defining gravitons, are not bound, but can cross into bulk space.

Given particle correlation by entanglement, the possible similarity of governing laws of different universes and the loop strings of M-theory, cross-world connectivity involving emulation by quasi-immanent causation cannot be claimed to be a nomological impossibility.

The Reversible Non-Cognitive Organism

A further nomological issue concerns the possibility of the existence of a reversible non-cognitive organism. This organism is metaphorically like living clay that can be molded to match a living human template by immanent causation. It appears to be the case that it is possible that such an organism could be generated in this universe; that is, our laws of nature do not preclude it. Research is moving in this direction and stem cell technology is already being employed to create human organs separate from a human body.[30] Further, as there are living human organisms in our universe, then the creation of one by technicians is possible; that is, the laws of nature do not preclude it, only the state of scientific progress to date.

Emulation

The resurrection possibility-model under consideration requires the Heavenly technicians to facilitate a body- and brain-copying process through immanent causation to preserve the identity of the ante-mortem person.

Body emulation does not carry so much identity baggage as brain emulation and above, I argue that the possibility of a reversible non-cognitive organism is not precluded by the laws of nature; both here and in Heaven if the natural laws are similar. Given this and the other cross-world possibilities referred to above, body emulation is not precluded and it survives the nomological criterion. However, whole brain emulation is a far more complex supposition.

As well as a profusion of science fiction films concerning mind uploading, in the domain of neuroscience there is current scientific research being undertaken on the possibility of emulating or copying the neural structure of a living brain to form an artificial electronic neural network that operates independently of the biological brain.[31]

The notion of mind-uploading is perhaps an inappropriate description of the process, giving the impression that mind is merely software that can be uploaded onto a computer and mental processes maneuvered like simple file transfers. However, although the brain is a physical entity like a computer, it is not a computer; neither does it operate like a computer.[32] Consequently, the term mind-uploading can create a false impression and a concomitant skepticism of the possibility of *mind-uploading*: whole brain emulation (WBE) is a more appropriate descriptor.

Current thinking suggests a gradual emulation will probably be the most successful form of WBE,[33] whereby an electronic neuron, comprising sensors (input) and actuators (output), is linked, possibly electromagnetically, to its biological counterpart. This synthetic neuron then synchronizes with its biological counterpart and, given time, becomes a working copy or emulation of it. The process is repeated until eventually synthetic neurons are synchronized with other synthetic neurons in a gradual emulation from a biological network to an electronic network. Given that there are approximately 86 billion neurons in the human brain,[34] this will be complex involving scanning, in-vivo nanotechnology and biotechnology.[35] This gradual process, it is argued, will preserve the phenomenal or mental continuity of the person so that the mind that emerges from the WBE is identical to the mind that emerges from the biological brain of the original person.[36]

Although successful research has been completed so far, effective WBE is some years off.[37] However, there is positive expectation:

> WBE is a deeply challenging and long-term prospect. Given current neuroscientific and technological knowledge there doesn't seem to exist any fundamental obstacles, merely a large amount of engineering and research [Sandberg, 2013, 14]

The WBE being considered at present is a synthetic emulation scenario, whereas the resurrection possibility-model is a biological emulation scenario. If synthetic emulation is possible, it seems that biological emulation is also possible given that the brain is biological; it may even be more straightforward. WBE is not precluded by the laws of nature of our universe and given the cross-world connectivity argument above, then cross-world WBE cannot be deemed to be a nomological impossibility.

From the above, I would argue that the proposition *Smith died and was resurrected* has cleared the hurdles of logical, conceptual, metaphysical, and nomological impossibilities, and as such its truth is a genuine possibility. However, Scott Sturgeon (2000) argues that only metaphysical and nomic possibilities mark genuine possibility; formal and conceptual possibilities do not.[38] Notwithstanding, I have included them in my argument as impossibility may have resulted from these forms.

The above analysis is only about *possibility* not *reality* and relates to a particular possibility-model. The *probability* of human resurrection is addressed in the following chapter.

CONCLUSION

In this chapter, I reflected upon the possibility of resurrection rather than the probability that resurrection actually takes place. This involved proposing a materialistic interpretation of resurrection as a process or model that could be considered as a genuine possibility. This approach could attract the charge of ad hoc wishful thinking, but I make no claim that such a possibility is an actuality, the intention is only to show that a materialist explanation of human resurrection is not an impossibility.

I began by placing God at the center of my resurrection hypothesis; that is, my hypothesis is dependent upon assumptions concerning the existence and will of God, although I acknowledge naturalistic possibilities in addition.

Given the nature of God outlined in Chapter 1, God, provided He exists, has the wherewithal to instantiate a resurrection scenario provided it is a possibility. Indeed, God can only instantiate that which is possible to instantiate and to establish such possibility I distinguished between conceivability and genuine possibility. I offered the proposed resurrection model up to Sturgeon's criteria of genuine possibility; logical, conceptual, metaphysical, and nomological criteria. The model adapted to the constraints of these criteria with the arguments of Chapter 6 influencing the adaptation. Finally, after considerable introspection under the lenses of the metaphysical and nomological criteria, I believe the model cleared Sturgeon's hurdles of genuine possibility.

Readers may find the idea of a Heaven-based, reversible non-cognitive organism capable of engaging in an emulation process somewhat hard to digest. Although the proposal is primarily design to fulfill the requirements of metaphysical possibility, I would remind the readers of traditional Christian and Judaic eschatology; the dead, despite organic annihilation, are raised incorruptible in the new age. Even the notion of a Cartesian soul that flutters off to Heaven really needs a new body to function as a person.[39] The proposal seems less incredible in the face of these resurrection hypotheses, although the atheist would no doubt dismiss the lot as palpable nonsense. What is of concern is the complexity of such a materialist possibility-model compared to a discarnate possibility-model;[40] complexity reduces probability.[41]

Pro forma possibility based upon a hypothetical model does not inspire confidence in a resurrection hypothesis; but once the necessary challenge of possibility is behind us, then the more challenging issue of probability looms forth. Undeniably, the force of evidence on the hypothesis as seen through the eyes of an atheist, agnostic, and theist is more telling and in the next chapter I engage with this probabilistic issue.

NOTES

1. Not all philosophers accept a notion of mind as being a property of brain states and therefore permanently bound to the physicality of the brain: Paterson R. W. K. (1995); Ch. 3.
2. Hauskeller M. (2016); p. 11.
3. Sturgeon S. (2000); Ch. 5.
4. Parfit D. (1987); Ch. 10; pp. 199–201; Parfit uses the expression teletransportation, whereas I use teleportation.

5. Xi-Lin Wang, Xin-Dong Cai, Zu-En Su, Ming-Cheng Chen, Dian Wu, Li Li, Nai-Le Liu, Chao-Yang Lu & Jian-Wei Pan (2015)—*Quantum teleportation of multiple degrees of freedom of a single photon*—in Nature 518; 26 February 2015; pp. 516–519.
6. Genuine Memory; Ch. 6.
7. Zimmerman D. (1997) for a more detailed analysis on immanent causation.
8. *Ibid*; p. 121.
9. Eric Olson (2010) prefers the descriptor 'ontic-leap model'; *ibid*; p. 62.
10. O'Connor T. & Jacobs J. D. (2010); p. 73.
11. Olson E. (2010); p. 62f.
12. Dilley F. B. (1977); p.35ff.
13. Corcoran refers to the simples as atoms; I prefer particles, but I continue with atoms for the sake of clarity.
14. The use of the word 'spooky' alludes to the letter sent by Albert Einstein to Max Born in 1947 concerning the quantum effect of entangled particles. He described it as *spukhafte Fernwirkung*; that is spooky action.
15. The Closer Continuer Theory; Ch. 6.
16. Loose J. (2013); p. 857f.
17. Baker L. R. (2000); pp. 39–41.
18. Ch.6; *The First-Person Perspective.*
19. Interestingly, Dean Zimmerman (2010) revisits his falling elevator model and engages with the constitution approach: Zimmerman D. (2010); p. 43 and Loose J. (2013), p. 863.
20. Loose J. (2013); p. 860.
21. Hick J. (1976); Ch. 15 for a comparable resurrection analysis.
22. Ch. 3; Sec. Two: The Teleological Argument; Cosmology and Multiverse Theory.
23. Colman S. & De Luccia F. (1980).
24. Greene B. (2011); pp. 155–161.
25. *Ibid*; Ch. 3.
26. Livio M. (2009); Ch. 9.
27. Superconductivity is a state of exactly zero electrical resistance in a conductor.
28. Blundell S. J. (2009)—*Superconductivity: A Very Short Introduction*—Oxford: Oxford University Press.
29. Clegg B. (2006).
30. Reardon S. (2015)—*New Type of Stem Cell Could Make it Easier to Grow Human Organs*—Scientific American; May 7 2015.
31. Sandberg A. & Bostrom N. (2008); p. 71.
32. Kurzwell R. (2005); pp. 149–153.
33. Sandberg A. & Bostrom N. (2008); p. 108.
34. Herculano-Houzel S. (2009); for a comparison of neuron size of the human brain capered to other mammals.
35. Sandberg A. & Bostrom N. (2008); pp. 41–52.
36. Cortese F. (2013); p. 4.
37. Sandberg A. & Bostrom N. (2008); pp. 98–100.
38. Sturgeon S. (2000); pp. 104–107.
39. Drange T. M. (2015); Ch. 12.
40. 1 Corinthians; 15:44.
41. Sudduth M. (2016); p. 208f.

Chapter 8

Home Straight

The Probability of Resurrection

My resurrection hypothesis has God at its center, hence Chapters 3, 4, and 5 where I established a range of values expressing the probable truth of the theistic proposition *God exists* given a variety of confirming and disconfirming evidential elements. The probability values established were hypothetical intersubjective values based upon consensus formation of a hypothetical betting syndicate and the judgement of two evaluators regarding the syndicate's expected decisions: vicarious probability.

Combined with the divine attributes established in Chapter 1, the theistic proposition has probabilistic force on the resurrection proposition and in this final chapter, I intend to establish the probable truth of that proposition in terms of a degree of belief. However, a theistic approach is not exhaustive; naturalistic resurrection hypotheses also exist and I intend to incorporate them into the analysis.

This chapter commences with an inter-probabilistic[1] methodology comprising a confirming interrelationship between three propositions:

1. p the theistic proposition *God exists*
2. r the resurrection proposition *dead persons will be resurrected*
3. R the Resurrection (of Jesus) proposition *Jesus died and was resurrected*

Background knowledge k is also an essential element in the propositional interrelationship, particularly related to naturalistic resurrection possibilities.

The analysis results in three probability values of r for the three members of the hypothetical syndicate. These three values then become the prior probability values for each member, prior, that is, to consideration of the two final evidential elements:

4. Near-death experience
5. Apparitions

As for Chapters 3, 4, and 5 the reader can also participate in the cumulative Bayesian process by keying in their own values as appropriate. Moreover, they are free to add additional confirming or disconfirming evidence they may consider relevant if they so wish (see Appendix 2). For example, they may wish to include *reincarnation* as a potential confirming evidential element, as did Ronald Paterson (1995) and Curt Ducasse (1961).[2] However, I have restricted my analysis to 4 and 5 above.

A NETWORK

I am employing two contributing propositions to the probable truth of the resurrection proposition r; namely, the theistic proposition p and the Resurrection (of Jesus) proposition R.

Prima facie, a Bayesian network[3] might apply to this tripartite analysis; that is, p to r; p to R then R to r. However, the application of Bayesian networking is *not* appropriate to this situation as the propositions are not probabilistically independent which is an essential feature of a Bayesian network. Consider for example the function $\Pr(R|p \;\&\; k)$, which is certainly greater than $\Pr(R|k)$, so the propositions R and p are not independent.

The Resurrection Proposition *R*

Despite this probabilistic dependence, the analysis is a network, albeit a non-Bayesian network; namely, the probable truth of r is affected by the truth of both p and R collectively and the truth of R is affected by p. The p-and-R relationship requires the following conditional probability values to be determined by the hypothetical syndicate: $\Pr(R|p \;\&\; k)$ and $\Pr(R|\neg p \;\&\; k)$.

The Resurrection of Jesus

The notion of Divine incarnation through Jesus suggests He was, *inter alia* human. If it was the case that Jesus was also crucified, died, and was resurrected, then I believe that such an extreme enigmatic event also has, to some degree, probabilistic confirming force on the human resurrection proposition *r*. In addition, the Gospels state that Jesus referred to human resurrection; for example, Matthew 22:30 and John 3:16–17.

At this point in the analysis, I am concerned with the conditional probability function $Pr(R|p \& k)$, where *R* is the proposition *Jesus died and was resurrected*, *p* the theistic proposition *God exists*, and *k* background knowledge. Thus, the function assumes God exists and it may be the case that He would want and have good reason to become incarnate on Earth at some point in the history of mankind to provide a lifestyle paradigm for all mankind. Jesus may have been the instantiation of that desire. Richard Swinburne (2003[A]) provides an argument why God, if He exists, would want this state of affairs,[4] but my immediate interest is in resurrection and for the sake of brevity I am assuming this incarnation in background knowledge *k*. Given such a divine incarnation, why would God want Jesus killed in the most appalling way only to be resurrected three days later to make His presence known to His disciples and others and then, over a month later, to ascend into Heaven?

The crucifixion of Jesus is theologically viewed as an atonement; that is, the self-sacrifice of a perfect life on the cross to atone for the sins of mankind. God accepts the sacrifice as a complete redress and reinstates the life lost by a bodily resurrection as confirmation of a debt paid. This sacrificial genre in the Gospels echoes the Jewish theology of the time, with God sharing the sacrifice with His people. This is representative of a divine endorsement of the life and work of Jesus as seen through first century Jewish culture. The Resurrection would be, in my terminology, a highly extreme enigmatic event,[5] what Swinburne terms a "super-miracle."[6] Thus, the supernatural force of the Resurrection can be seen as giving further authority to Jesus' way of life and teaching, particularly the forgiveness of sin; a divine endorsement and exaltation of Jesus. Moreover, the Resurrection was the seed that spread the growth of the Gospel and Jesus as the forerunner of human resurrection. In short, given the existence of God (*p*) and His incarnation in Jesus (in *k*), the Crucifixion, Resurrection, and Ascension were essential self-sacrificial and enigmatic events that leave little doubt as to God's vision for the ideal way of life for mankind, as well as providing hope for our own resurrection.

Counter Argument

In this part of the argument, God is assumed to exist and is incarnate in Jesus with the intention of providing an ideal way of life model for mankind. Rather than permitting Himself to be viciously beaten and crucified, Jesus could have employed His divine power to create an extreme enigmatic event to persuade people of His divine status and preferred way of life; pain, death, and resurrection were simply unnecessary events for fulfilling God's intention.

Reply

Sacrifice as an atonement for sin was a central feature of Jewish theology of that time; Jesus' sacrifice for the atonement of sins would have engaged with such a world-view, exalting His mission in that cultural context. Moreover, the notion of bearing the sin of others through pain and sacrifice so that they are forgiven has the hall-mark of selfless love; a further facet of God's desired relationship with mankind as well as His desire for the relationship between people. Without death, there could be no resurrection, and although there were other reported extreme enigmatic events in Jesus' life such as walking on water, turning water into wine, curing the blind, resurrecting Lazarus, and others, His own resurrection and appearances would bear the mark of Divine Will, providing authority to His overall mission. Moreover, Jesus' reported resurrection would be indicative of God's vision for bodily human resurrection. This is a case of synergy; individually the events and objectives seem exchangeable for a more benign approach, collectively they provide a powerful Divine statement to motivate people to continue Jesus' mission into the distant future which is, in fact, what happened.

Swinburne's Probabilistic Argument

Richard Swinburne (2003^A) provides an inter-probabilistic Bayesian argument supporting the Resurrection proposition through the proposition *God became incarnate* (c).[7] However, he selected variables that I consider questionable; that is, 0.5 for both $Pr(t|k)$ and $Pr(c|t\ \&\ k)$ seemingly on the basis of the principle of indifference, where t is the theistic proposition.[8] I am reticent about the use of this principle;[9] William Rowe (1996)^A also selected, incorrectly I believe, this value for the prior probability of his theistic proposition in his argument from evil.[10] Likewise, Curt Ducasse (1961) also applied the principle of indifference to his resurrection analysis.[11] Moreover, 0.5 for $Pr(t|k)$ is too high for an atheist and too low for a theist, distorting the analysis. Swinburne also selects a value of 0.1 for $Pr(f|c\ \&\ k)$ where f is the evidence surrounding this incarnation; that is, a unique prophet and resultant super-miracle.[12] Swinburne believes this value to be too low[13] and I agree; notwithstanding, it is left at this value. His calculations conclude with $Pr(c|f\ \&\ k) = 0.97$ and on the basis of this result Swinburne argues that it is "very probable indeed that God became incarnate in Jesus Christ who rose from the dead." [Swinburne, 2003^A, 214]

EVALUATORS' CONCLUSION

Assuming God exists (p) and that Jesus lived and pursued a mission as presented in the Biblical Gospels, then $\Pr(R|p \,\&\, k) >> \Pr(R|\neg p \,\&\, k)$, as it is extremely unlikely that Jesus would be resurrected if God did not exist and the above argument for God's desire for the Resurrection is persuasive (provided He exists).[14] Indeed, $\Pr(R|\neg p \,\&\, k)$ must be extremely low. The only alternative is that a supernatural being with similar, but less forceful, attributes to the omni-predicates could cause the Resurrection. Moreover, the naturalistic resurrection possibilities addressed below are futuristic and therefore do not apply to this historical event.

Evaluators' Judgement

The hypothetical syndicate would be inclined to agree that $\Pr(R|p \,\&\, k) >> \Pr(R|\neg p \,\&\, k)$ at the theoretical level.

Consensus

The evaluators believe that the hypothetical syndicate would form a consensus on the probability value of the conditional as follows:

$\Pr(R|p \,\&\, k) = 0.7$
$\Pr(R|\neg p \,\&\, k) = 0.01$

The low value for $\Pr(R|\neg p \,\&\, k)$ assumes that exotic naturalistic auxiliary assumptions for the Resurrection in k are unjustifiable. (See Appendix 1)

The Person-Resurrection Proposition *r*

It would be possible to allocate a probability-value to $\Pr(r|k)$ for the three members of the hypothetical syndicate, although given possible naturalistic explanations and variations in theistic belief it would be very difficult.[15] Thus, I have chosen to establish $\Pr(r|k)$ by way of the conjunction of the theistic and Resurrection propositions ($p \,\&\, R$) through a probability network.

This is the final stage of this simple network. However, the normal use of the theorem of total probability is limited; that is, $\Pr(p \,\&\, R \,\&\, k)$ cannot be expanded to $\Pr(p|k)\Pr(R|k)$ as is usually the case with the total probability function. This is due to the lack of probabilistic independence of the variables and the following equivalent form of the theorem of total probability must be employed:

$$\Pr(r|k) = \Pr(r|(p \,\&\, R) \,\&\, k)\Pr((p \,\&\, R) \,\&\, k) + \Pr(r|\neg(p \,\&\, R) \,\&\, k)\Pr(\neg(p \,\&\, R) \,\&\, k)$$

Prima facie, the constituent functions of the theorem appear difficult to determine. However, they can be changed to simplify the determination of their values as follows (k is excluded for purposes of clarity only):

1. *From the theory of total probability*:
   $\Pr(r) = \Pr(r|p \,\&\, R)\Pr(p \,\&\, R) + \Pr(r|\neg(p \,\&\, R))\Pr(\neg(p \,\&\, R))$
2. *From 1*:
   $\Pr(r) = \Pr(r|p \,\&\, R)\Pr(p \,\&\, R) + \Pr(r|\neg p \,\&\, R)\Pr(\neg p \,\&\, R) + \Pr(r|p \,\&\, \neg R)$ $\Pr(p \,\&\, \neg R) + \Pr(r|\neg p \,\&\, \neg R)\Pr(\neg p \,\&\, \neg R)$
3. *From Df conditional* probability: $\Pr(p \,\&\, R) = \Pr(R|p)\Pr(p)$
   From 2 & 3:
   $\Pr(r) = \Pr(r|p \,\&\, R)\Pr(R|p)\Pr(p) + \Pr(r|\neg p \,\&\, R)\Pr(R|\neg p)\Pr(\neg p) + \Pr(r|p \,\&\, \neg R)\Pr(\neg R|p)\Pr(p) + \Pr(r|\neg p \,\&\, \neg R)\Pr(\neg R|\neg p)\Pr(\neg p)$

There are three propositions that constitute the variables in this total probability formulation r, R, and p with k as background knowledge (not shown for clarity). Determining $\Pr(r|k)$ is the objective and R has been considered above. This leaves p and k and there are two considerations relevant to them: the will of God implied by p and naturalistic explanations for human resurrection residing in k, and I address these considerations as follows:

The Will of God

In Chapter 1, I argue that the essential attributes of God are that of omnipotence, omniscience, omni-benevolence, and perfect freedom and given the possibility of human resurrection established in Chapter 7, these attributes would ensure the potential for a resurrection scenario provided it is the will of God. This is particularly relevant when faced with the issue of divine justification associated with the problem of evil introduced in Chapter 4. The greater good derived from human free-will can lead to horrendous suffering, but this can be mitigated by a divine resurrection strategy adding to the greater good. Assuming God exists, it is reasonable to assume that such a strategy would be employed by Him. However, divine action is not the only support for a resurrection hypothesis; naturalistic scenarios are also possible.

Naturalistic Possibilities

Within the background knowledge k there are futuristic possibilities for person-resurrection that possess probabilistic confirming force upon r.

Cryonics and Emulation

Cryonics is a life preserving system whereby humans are immediately frozen upon death and held in that state pending medical developments whereupon they are defrosted, revived/cured if necessary and resurrected. Although this process is only available to those who can afford to join a cryonics procedure, it is technically a resurrection possibility. Likewise, for the body and brain emulation process introduced in Chapter 7. The exchange of biological body parts for synthetic substitutes may ensure a very long 'life' for the individual, but this is a life extension rather than a resurrection scenario. However, combined with the cryonics process, resurrection may be realized. At present cryonics is a limited and unrealized service, but if shown to be successful, particularly in combination with synthetic emulation, it could be rolled-out to a bigger population and the truth of *r* fulfilled.

Counter Argument

It could be argued that a frozen suspension is not death, merely a cessation of body and mental functions for a period, rather like a near-death state of affairs where a patient's heart stops beating and neural functions cease; however, the patient is revived by medics. Death really means 'beyond the brink'; that is, not revivable. Thus, the concept of *resurrection* may not apply to the cryonics process; only *revivable*.

Reply

The notion of *reviving* is stretched in a cryonics scenario because of the duration involved and the cessation of all bodily functions. The patient, is effectively dead and bodily decomposition stemmed by freezing. It seems that both the concepts of *revival* and *resurrection* have applicability in such a context.

A Digital Afterlife

Eric Steinhart (2014) offers a detailed futuristic emulation possibility-model involving the digitization of peoples' cellular structure and chemistry to such a degree of detail that they are effectively alive and conscious in a digital form; that is, their entire body, cells, and chemistry 'live' in a computerized afterlife with their old biological defects digitally rectified.[16] Such emulation extends to an engineered virtual environment (the terrarium),[17] including its wildlife to such degree of precision that an electronic-life, in terms of a machine, is equivalent to a biological-life, equally argues Steinhart, in terms of a machine.[18] This echoes the imaginary teleportation example introduced in Chapter 7, where I argue that teleportation does not preserve identity.[19] Steinhart evades such diachronic person-identity challenges by denying it exists; that is, person-identity cannot be preserved over time, no matter how short the time.[20] Thus, with Steinhart's model, young-Smith is not identical to young-Smith plus a minute and certainly not old-Smith. Given this loss of identity over time, then a digital afterlife is no different from getting older in terms of persistence. However, this is not to deny continuity; it is a case of electronic life after death, not biological life through death.[21] Indeed, Steinhart asserts that resurrection, as life through death in the way I portray it in this book, is illogical [22] and or impossible.[23]

Counter Argument

Steinhart's position on person-identity is too strict; that is, if all the properties of Smith at t_1 are the same as all the properties of Smith at t_2, then Smith at t_1 is identical to Smith at t_2.[24]

Steinhart would argue that, because of all the changes occurring in Smith at the biological and molecular levels, such diachronic identity fails. Moreover, Steinhart's life uploading involves third-party intervention, a breach of the only x and y principle of identity theory.[25] There are issues concerning diachronic person-identity which I address in Chapter 6; however, we do not live our lives in any domain that denies our identity over time, be it personal, social, moral, legal, or scientific to name but a few.

Regarding Steinhart's claim that biological resurrection is illogical and/or impossible, I refer the reader Chapter 7: Genuine Possibility.

Reply

I do not think Steinhart's model necessarily fails because of identity-denial; piecemeal digital emulation may well satisfy the criteria diachronic person-identity.[26] However, the only x and y principle remains a potential problem as it does in my argument for resurrection.

Interestingly, the possible amelioration of human suffering through resurrection presented in Chapter 1 is also a potential in Steinhart's model. [27] Moreover, Steinhart's model exhibits a beguiling parallel to my resurrection possibility-model of Chapter 7, albeit digital rather than biological.[28] Moreover, assuming we are already in a simulated digital universe, his model includes a deity; an ultimate simulation engine that has instantiated a finite series of nested digital simulations. Each simulation has created the next simulation along the chain that nests in it, whose digital-civilization parallels the maximal attributes of the simulation-deity, but sequentially diminishing in quality the further from the deity each simulation is. Like John Hicks' soul-making theodicy, an individual's digital-progress towards the deity follows from their personal performance in each simulation of the series.[29]

Omega Point Theory

A rather bewildering resurrection possibility has been proposed by Frank Tippler (1994) termed the Omega Point Theory, a distant future evolutionary analysis of the universe.

Omega Point was first posited by Pierre Teilhard de Chardin in 1950; he argued that the universe will eventually attain maximal complexity, and consciousness will be omni-present with mankind attaining extreme levels of consciousness. Building on this notion, together with the assumption that the universe is a closed system,[30] Tipler argues that, in the distant future, conscious life will saturate the universe and that sentient beings will attain knowledge and understanding far beyond the present state of affairs. Like Eric Steinhart (2014), he argues that information technology, in the form of artificial intelligence, will be at the heart of such knowledge and understanding, to the point where e-simulations of the universe and all events, past, present, and future, will be emulated.[31] This emulation will reach its zenith (omniscience) when the universe collapses back to a singularity. It is this simulation that represents Tipler's resurrection hypothesis; that is, the dead will be raised as an intrinsic part of the simulation. Despite the eventual termination of the universe, simulated existence would continue indefinitely. The theory is analogous to that of Biblical eschatology.[32]

Counter Argument

The most potent defeater of Tipler's Omega Point Theory was discovered by the Supernova Cosmology Project team led by Saul Perlmutter in 1998. The project set-out to measure the de-acceleration of the universe required for an eventual collapse as predicted by Tippler and most astronomers. However, much to the consternation of the astronomy fraternity, they discovered the opposite; the expansion of the universe is accelerating,[33] the cause was termed *dark energy*, an enigmatic force counteracting the force of gravity within the universe. For this discovery, Saul Permutter was awarded the Nobel Prize for physics in 2011.

I argue in Chapters 6 and 7, that person-identity must be preserved in a resurrection scenario; that is, the following criteria must be satisfied:

1. A unique being; that is, not simultaneously duplicated.
2. Possessing a mental/noetic system.
3. Possessing a physical body.
4. Psycho-physical congruity in continuity; that is, no physical or psychological branching in the existence time-line.
5. Immanent causation; that is, diachronic identity is maintained by intrinsic causal factors.
6. Satisfying the only x and y principle; that is, "…whether a later individual y is identical with an earlier individual x can depend only on facts about x and y and the relationships between them: it cannot depend upon facts about any individuals other than x and y." [Noonan, 2003, 129]

Tippler's resurrection hypothesis satisfies criteria 1, 2, and possibly 6, but criteria 3, 4, and 5 are breached. Perhaps criteria 3 and 4 can be overlooked as all persons are simulated in Tippler's model as are their physicality; thus, there will be no appearance of physical or mental branching. However, criterion 5 is a serious breach, as the simulated resurrection has been caused by extrinsic factors (artificial intelligence) to the deceased person. The simulated person is not the same person who died eons ago as a material organism. Extending the person-identity criteria to include *identity by constitution*[34] and the *first-person perspective*,[35] the Tippler model also falters on these points.

Androids

If in the future, androids become conscious and attain the status of personhood, but occasionally break down and are repaired to full functionality, this may be a case of person-resurrection as defined by the proposition *r*; that is, *dead persons will be resurrected*. Whether the concept of *death* would apply to an android is questionable; *broken-down* seems more applicable. However, it may become applicable if the existence of androids as persons becomes established. This scenario may be combinable with human ante-mortem synthetic whole brain emulation.

EVALUATORS' CONCLUSION

To determine the probability of *r*, the component functions of the theory of total probability need to be allocated values; that is, the following needs to be evaluated:

$\Pr(r) = \Pr(r|p \,\&\, R)\Pr(R|p)\Pr(p) + \Pr(r|\neg p \,\&\, R)\Pr(R|\neg p)\Pr(\neg p) + \Pr(r|p \,\&\, \neg R)\Pr(\neg R|p)\Pr(p) + \Pr(r|\neg p \,\&\, \neg R)\Pr(\neg R|\neg p)\Pr(\neg p)$ (background knowledge *k* excluded for clarity):

The following component functions are already known:

1. $\Pr(R|p \,\&\, k)$, which has been determined by the hypothetical syndicate above.
2. The prior probability function $\Pr(p|k)$ which has been determined by the hypothetical syndicate through Chapters 3, 4, and 5.

Thus, the remaining unknown component functions required are: $\Pr(r|p \,\&\, R \,\&\, k)$; $\Pr(r|\neg p \,\&\, R \,\&\, k)$; $\Pr(r|p \,\&\, \neg R \,\&\, k)$; $\Pr(r|\neg p \,\&\, \neg R \,\&\, k)$

Aggregation

In Chapter 5, I introduced Loren Meierding's (1998) *consensus gentium* argument, where he presented an aggregation argument for the probability of the theistic proposition based upon the evidence of more people believing in God than disbelieving. He reflected on the likelihoods $\Pr(e|p \,\&\, k)$ and $\Pr(e|\neg p \,\&\, k)$,[36] arguing that the former contains both divine and naturalistic force on the truth of the evidence rather than the naturalistic force only of the latter. The evidence e signifies more people believing in God than disbelieving, the theistic proposition being p and naturalistic arguments for e are within k. On the basis of the conjunction of p and k in the former function, he concluded that $\Pr(e|p \,\&\, k) > \Pr(e|\neg p \,\&\, k)$. The same form of aggregation argument applies to the functions $\Pr(r|p \,\&\, R \,\&\, k)$, $\Pr(r|\neg p \,\&\, R \,\&\, k)$, $\Pr(r|p \,\&\, \neg R \,\&\, k)$, and $\Pr(r|\neg p \,\&\, \neg R \,\&\, k)$.

Counter Argument

There is the possibility of over determination;[37] for example, the probability of the gazelle drinking from the pool if a cheetah is present would be the same if a cheetah and a hyena were present. With the case in point, does R and p ensure an increase in confirmatory force on r compared to just R or just p?

Reply

The confirmatory force of R upon r is a result of Jesus' humanity; that is, the person in the form of Jesus was resurrected, thereby providing (assuming its truth) an example of person-resurrection. The confirmatory force of p upon r is a result of the will of God combined with the wherewithal to ensure person-resurrection: over determination does not apply to this scenario.

Evaluators' Judgement

Given the naturalistic possibilities for r residing in k, there is an aggregate effect; that is, assuming no over determination, r is more probable the greater number of confirming variables. Therefore, $\Pr(r|p \,\&\, R \,\&\, k) > [\Pr(r|\neg p \,\&\, R \,\&\, k)$ and $\Pr(r|p \,\&\, \neg R \,\&\, k)] > \Pr(r|\neg p \,\&\, \neg R \,\&\, k)$. The evaluators also believe that the syndicate would consider that $\Pr(r|p \,\&\, \neg R \,\&\, k) > \Pr(r|\neg p \,\&\, R \,\&\, k)$.

Consensus

Consider the following:

1. $\Pr(r|p \,\&\, R \,\&\, k)$: in this function God is assumed to exist, Jesus existed and was resurrected and the naturalistic possibilities reside in k. Given Jesus is an example of human resurrection, albeit possibly a divine one, and the implications of the will of God implied by p augmented by the naturalistic possibilities for resurrection, then $\Pr(r|p \,\&\, R \,\&\, k)$ is very high. The

evaluators believe that the syndicate would accept the following for this posterior probability function:
$\Pr(r|p \,\&\, R \,\&\, k) = 0.9$

2. $\Pr(r|p \,\&\, \neg R \,\&\, k)$: given the assumptions of this function, a high value is also applicable; less than $\Pr(r|p \,\&\, R \,\&\, k)$ without the assumption of the Resurrection proposition R, but given p & k the evaluators believe the syndicate would accept the following:
$\Pr(r|p \,\&\, \neg R \,\&\, k) = 0.7$
3. $\Pr(r|\neg p \,\&\, R \,\&\, k)$: this function is singular as it assumes Jesus was resurrected yet God does not exist. Given God is defined by the omni-predicates of omnipotence, omniscience, omni-benevolence, and perfect freedom, it is possible that a powerful supernatural being without these maximal attributes (a case of $\neg p$) could have resurrected Jesus. Despite the probability of such a state of affairs being low, the Resurrection is still assumed true in this function and is an example of human resurrection. Moreover, naturalistic explanations still reside in k augmenting the confirming force of R and the evaluators believe the syndicate would accept the following:
$\Pr(r|\neg p \,\&\, R \,\&\, k) = 0.6$
4. $\Pr(r|\neg p \,\&\, \neg R \,\&\, k)$: the confirming force of this function is limited to the naturalistic possibilities for resurrection residing in k; nevertheless, the naturalistic possibilities are not without reasonable confirming potential and the evaluators believe the syndicate would accept the following:
$\Pr(r|\neg p \,\&\, \neg R \,\&\, k) = 0.5$

In addition to the hypotheses p and R and their negations, it should be noted that the above values also depend upon the auxiliary assumptions residing in k. (See Appendix 1) These can be unreasonably bulked to support r in the likelihoods, but they should be justified for the likelihood to be well-defined. The values selected are based upon justified auxiliary assumptions that are embedded in the argument. (See Appendix 1)

Bayesian Conditionalization

The two variables p and R require existential assessment for conditionalization to occur. The proposition p was assessed in Chapters 3, 4, and 5 leaving the proposition R to be considered.

The evidence of Jesus' resurrection is an important requirement in this stage of the analysis to give substance to the inclusion of R as evidence in the posterior function $\Pr(r|p \,\&\, R \,\&\, k)$. Space limits detail, but the following outlines the variety and sequence of Biblical testimony of Jesus' post-resurrection appearances:

1. To Mary Magdalene close to Jesus' tomb (Matthew 28; Marcan appendix[38] and John 20)
   1.1. To Mary, the mother of James 'the other Mary,' Salome, Joanna, and 'other women' (Matthew 28; Mark 16; Luke 24)
2. To Peter (Luke 24 and 1 Corinthians 15)
3. To Cleopas and others along the Road to Emmaus (Luke 24)
4. To the eleven (less Thomas) (1 Corinthians 15,[39] Luke 24, and John 20)
5. To the eleven (incl. Thomas) (John 20)
6. Seven disciples at the Sea-of-Tiberias (John 21)
7. To a large gathering (500) on a Galilean mountain (Matthew 28; 1 Corinthians 15)
8. Meeting his half-brother James (1 Corinthians 15)
9. To the disciples at the Mount of Olives (Luke 24; Acts 1)
10. To Saul (subsequently Paul) on the road to Damascus (Acts 9; 22)[40]

Swinburne reconciles the accounts to specific locations: 1/1.1, 2, 3, 4 as different locations; 5, 6, 7, 8 to the Galilee location; 9 to Jerusalem and 10 on the road to Damascus.[41] He also acknowledges minor errors in the Gospel accounts due to the nature of oral testimony and theologizing, but the essential history is sound.[42]

Counter Argument

The first gospel by Mark was written forty years after Jesus' death and over that time the Jesus story could evolve into a legendary rather than historical account, particularly given the oral tradition of recounting events. The gospels could also have been subject to modification and, given the theological content, deliberate false testimony cannot be dismissed. Moreover, the gospels were written in Greek and liable to be influenced by Greek culture, with their miracle myths and multiple gods. Indeed, resurrection myths are common from regions close to Israel and may have influenced the gospel writers: Osirus in Eygpt (c. 2000 BCE); Tammuz a Sumerian god (c. 2000 BCE); Dionysus a Greek god (c.1200 BCE); Adonis a Greek god (c.600 BCE); Attis a god from Asia Minor (c.200 BCE).

Reply

Analyses of the historicity of the Resurrection of Jesus (the Resurrection) are plentiful, but space limits me to a brief reply to the above challenge.

There is considerable evidence surrounding the Resurrection hypothesis and, like all explanations of evidence, there are multiple explanatory possibilities. A common process when faced with such variety, is to select one by inference to the best explanation. Listed below is relevant evidence of the Resurrection and from this and the above, the evaluators will decide whether the hypothetical syndicate would adopt the probability values of *R* above.

1. The empty tomb testimony:
   1.1. The Resurrection was first announced in Jerusalem where Jesus was buried; a risqué strategy if His body was still in the tomb. Moreover, the first empty tomb testimony was by women (see 1 above), female testimony was not valued at that time in that culture; a legendary testimony would be expected to be male testimony.[43]
   1.2. Joseph of Arimethea, a member of the Jewish Sanhedrin, buried Jesus [44] and then later, the Jewish opposition to Christianity of that time acknowledged the empty tomb; that is, the stolen body explanation.[45] This is a hostile source confirming the post-burial empty tomb, albeit with an alternative explanation.
   1.3. The pre-Markan source[46] is dated to a few years from the event; weakening the above evolving story explanation regarding the empty tomb.
   1.4. Mark's explanation is very simple compared to a typical legendary explanation such as the apocryphal gospel of Peter, which elaborates the burial and Resurrection.[47]
2. The appearances:
   2.1. The earliest appearance creed is in Paul's 1 Corinthians 15:3–8, considered to be from Peter and James, who witnessed the risen Christ and testified this to Paul three to five years after the Crucifixion. This is a close-to-the-event eyewitness testimony conveyed directly to Paul as opposed to being handed down through many intermediaries.[48]
   2.2. Deliberate deceit seems implausible given that the disciples were prepared to die for their beliefs and self-delusion is equally implausible given that the risen Christ was a physical body not an apparition. Thus, it is reasonable to assert that the Disciples believed the physical risen Christ appeared to them. Perhaps they were hallucinating; however, touching and socializing is hardly the evidence of mass hallucination.

The Resurrection is not an undisputed fact and the atheist would be quite happy to adopt a lower value for $\Pr(r|(p \,\&\, R) \,\&\, k)$ than the theist (agnostic) and theist, but would still acknowledge that $\Pr(r|((p \,\&\, R) \,\&\, k) > \Pr(r|((\neg p \,\&\, \neg R) \,\&\, k)$. All three members of the hypothetical syndicate would accept the naturalistic possibilities within *k* and would be happy to adopt $\Pr(r|((\neg p \,\&\, \neg R) \,\&\, k) = 0.4$. However, although the evidence would probably convince the theist (agnostic) and theist to hold the value $\Pr(r|((p \,\&\, R) \,\&\, k) = 0.9$, the evaluators believe that the atheist, based on the evidence, would prefer the value $\Pr(r|((p \& R) \& k) = 0.7$.

Prior Values for Pr(*r* | *k*)

All the probability values are now available to determine $\Pr(r|k)$ from this network and are presented in the following tables:

Table 8.1. The Proposition *R* on the Proposition *p*

	¬p	p
¬R	0.99	0.3
R	0.01	0.7

This is a conditional probability table indicating the probability values of the propositions R and $\neg R$ assuming the truth the propositions p and $\neg p$ respectively. Note that the two value-columns each sum to 1, as the disjunction R or $\neg R$ is certain.

Table 8.2. The Proposition *r* on the Proposition *R* and Proposition *p*

	p		¬p	
	¬R	R	¬R	R
¬r	0.30	0.10 (0.3)	0.50	0.40
r	0.70	0.90 (0.7)	0.50	0.60

Atheist's value in parenthesis

This is also a conditional probability table indicating the probability values of the propositions r and $\neg r$ assuming the truth the proposition conjunctions $p \& R$, $p \& \neg R$, $\neg p \& R$ and $\neg p \& \neg R$ respectively. Note again, that each of the four columns sum to 1 for the reason given in Table 8.1 and that an additional probability-value (in parenthesis) has been included for the atheist.

Table 8.3. The Theistic Proposition *p*

	Atheist	Theist (agnostic)	Theist
Pr(p|k)	0.10	0.71	0.93
Pr(¬p|k)	0.90	0.29	0.07

This table indicates the prior probability values for the theistic proposition p for the three hypothetical syndicate members determined from Chapters 3, 4, and 5. The value columns sum to 1 because the negation of the theistic proposition $\neg p$ has also been included.

From the theorem of total probability:

$\Pr(r) = \Pr(r|p \,\&\, R)\Pr(R|p)\Pr(p) + \Pr(r|\neg p \,\&\, R)\Pr(R|\neg p)\Pr(\neg p) + \Pr(r|p \,\&\, \neg R)\Pr(\neg R|p)\Pr(p) + \Pr(r|\neg p \,\&\, \neg R)\Pr(\neg R|\neg p)\Pr(\neg p)$ (k has been left out for purposes of clarity only)

From the theorem of total probability and Tables 8.1, 8.2, and 8.3:

Atheist

$\Pr(r) = (0.70 \times 0.70 \times 0.10) + (0.60 \times 0.01 \times 0.90) + (0.70 \times 0.30 \times 0.10) + (0.50 \times 0.99 \times 0.90) = 0.52$

Theist (agnostic)

$\Pr(r) = (0.90 \times 0.70 \times 0.71) + (0.60 \times 0.01 \times 0.29) + (0.70 \times 0.30 \times 0.71) + (0.50 \times 0.99 \times 0.29) = 0.74$

Theist

$\Pr(r) = (0.90 \times 0.70 \times 0.93) + (0.60 \times 0.01 \times 0.07) + (0.70 \times 0.30 \times 0.93) + (0.50 \times 0.99 \times 0.07) = 0.82$

NEAR-DEATH EXPERIENCE

A near-death experience (NDE) describes a *close to the brink of death*, enigmatic experience that is recalled by the experiencer upon recovery.[49] Descriptions are of a heavenly realm visited by the experiencer. Many publications have emerged on the subject, particularly from those who have had NDEs. [50]

Commonality of Testimony

NDEs are unavoidably testimony based and, despite some descriptive variations, the testimonies have some common characteristics and Raymond Moody (2001) provides details of such common characteristics:[51]

1. *Hearing the news*: despite being in a near-death state the patient can hear the medics pronouncing them dead or close to death.

2. *Feelings of peace and quiet*: despite prior suffering the patients sense a feeling of painless serenity.
3. *The noise*: the NDE patient hears clicking, buzzing, whistling, or humming noises.
4. *The dark tunnel*: a classical experience, but not every NDE patient experiences it; that is, moving towards light through a dark tunnel.
5. *Out of the body*: this is the experience of leaving the near-death body and floating above and around it observing events taking place close by. Sometimes the reported events are confirmed by third parties as true.[52]
6. *Meeting others*: meeting and conversing with related and unrelated dead persons.
7. *The being of light*: meeting and communicating with an endearing person radiating bright white light who questions the patient's readiness to die.
8. *The review*: the being-of-white-light engages the patient in a review of their life to invoke a personal reflection.
9. *The border or limit*: some, but not all testimonies include a point-of-no-return border of various types with the understanding that to cross the border entails no return to the patient's world.

Meeting and conversing with related and unrelated dead persons is of particular interest. If the resurrection hypothesis is true, then it would be reasonable to assume that some resurrected relatives and their new friends would want to greet the new arrival. This is an auxiliary assumption (see Appendix 1) and is no different to human behavior in our world. Indeed, it is normal for most people to want to see their loved ones and relatives if they have been separated for some time.

The Return

10. *Coming back*: there is a reluctance to return because of the attractiveness of this other world, but a sense of obligation and responsibility to others draws them back.
11. *Telling others*: so vivid and enlightening are the experiences, there is a desire to testify to the experience coupled with a reluctance to speak-out so as to avoid ridicule.
12. *Ineffability*: difficulty finding the language to describe the NDE.
13. *Effects on lives*: a more reflective state of mind results, with a greater empathy for others.
14. *New views on death*: a rejection of the annihilation view of death.

The above pattern is generally repeated from case to case and the similarity is indeed a beguiling state of affairs; why should the descriptions be so similar given that there is no contact between the different experiencers? Although the descriptions may be embellished with contingent personal and cultural minutia, the essential content is remarkably similar.

Persons who have had NDEs claim they no longer fear death and undergo positive psychological changes, including a greater empathy for others.[53] Not all NDE reports are pleasant and some are of terrifying Dante-like experiences. Out-of-body experience testimony often accompanies NDE testimony; that is, a claim that the perceptual consciousness of the experiencer leaves and hovers over their body and moves around observing their environment before going into a NDE state.[54]

One perplexing post-NDE phenomenon reported by Penny Sartori (2014) is that of the failure of mechanical devices, particularly watches, on or near to an ex-NDE patient. Although some testimonies are quoted, no data is provided on the numbers who experience this phenomenon.[55] Also perplexing, are those cases where an NDE patient learns new information from personal encounters during their NDE episode; for example, meeting a friend or relative who has recently died, but where their death was unknown to the patient prior to the NDE episode.[56]

If it is the case that there is immediate human resurrection upon death, then NDEs may be an experiential intermediary stage of the resurrection process. It is a process that can result in the death of the person in this world and their resurrection in another, or the aborting of the dying process and a reawakening in this world. Thus, evidence of NDEs may have probabilistic confirming force on the resurrection proposition *r*.

Subjectivity

One insurmountable problem with NDE testimony is that it is entirely subjective and, in that sense, unchallengeable and because of this, NDE testimony gains traction. If I claim I am in pain then, despite a plethora of medical tests that cannot find a source of the pain, it cannot be refuted. Unless all the NDE claimants are colluding in a grand deception, then it is clear that NDEs occur. However, to then claim they are a veridical experience of an alternative reality is another matter. This makes such experiences one removed from pain experiences. At least with pain I can frequently point to a cause such as an injury. The experience of an alternative reality is an unverifiable assumption. This gives NDE testimonies even further psychological traction to the point where some believe they are proof of Heaven.[57]

Within such a highly subjective context, testimony can easily become embellished to convey a strength of conviction by the experiencer and perhaps to impress the listener, as may be the case with children's NDEs.[58] Unintentional and intentional embellishment[59] could have the effect of obscuring potential materialist explanations. All that can be provided in this highly subjective domain are alternative causal explanations and it is up to, in this case, the hypothetical syndicate to make an inference to the best explanation.

Timing Issues

Michael Marsh (2010) asserts that there is no supernatural causation of the experience; it is neurological. He argues that although the NDE appears to the experiencer to occur over a long duration, it is a short-term brain event occurring just prior to recovery or re-awakening from the near-death state. By way of evidence he employs the word-counting methodology used by dream researchers to estimate the duration of a person's dream; the less the number of words describing the dream, the shorter the dream and NDE descriptions tend to be short.[60] He also refers to the NDE-like descriptions of unsuccessful suicide attempts from the Golden Gate Bridge; a four second drop, but often for the survivor an imagined experience of a long duration.[61] Moreover, pilots who blackout in a training centrifuge machine also experience NDE-like mental events over a seemly long period; however, the unconscious state is less than one minute.[62] Thus, what seems to the experiencer to be a long sojourn into the hereafter, is but a few moments just prior to re-awakening.

John Fischer and Benjamin Mitchell-Yellin (2016) also consider the timing of NDEs, suggesting that reported experiences may not actually occur at the time and for the duration that the experiencer thought they did. From this they argue that alternative explanations arise to that of supernatural ones; for example, vague perception of events prior to and after cessation of normal brain activity. The experiencer upon recovery, then mentally projects the timing of the experiences to the period when the brain was unconscious. Similarly, remembered events can be recalled in the semi-awakened state and likewise projected to the unconscious period.[63]

Reply

To claim that, because there are some inaccurate timing projections, all NDEs are a consequence of these inaccuracies is not justified. Nevertheless, the evidence does provide some probabilistic support for a naturalistic explanation of NDEs. With likelihood comparisons such as $\Pr(NDE|N\ \&\ k)$ to $\Pr(NDE|S\ \&\ k)$, where N is the naturalist timing-hypothesis and S a supernatural hypothesis, much depends upon the auxiliary assumptions adopted; for example, *an afterlife is fictitious or real*. It is the justification of these assumptions which makes the difference, and additional evidence and argument bear upon this, as well as the receptiveness of the analyst in terms of their prior beliefs.

Neurological, Physiological, and Cultural Explanations

Michael Marsh (2010) provides a detailed neurological explanation for NDEs.[64] Of particular interest is NDE testimony involving the shunning of worldly moral responsibility at the outset of the NDE to the return of such responsibilities towards the end of the NDE and re-awakening. He suggests the first phase is due to a cessation of frontal cortex function and the second phase a re-engagement of frontal cortex function.[65] He suggests this process is due to oxygen depletion (cerebral hypoxia) at the outset and re-oxygenation upon recovery.

Marsh also reviews NDE testimonies, concluding that they are idiosyncratic, geo-centered descriptions orchestrated around the experiencer's prior conception of death, the afterlife, and their religious beliefs rather than new discoveries. The descriptions are predominantly in the first person and simple. Marsh believes this is indicative of subconscious memory activity rather than perception by a disembodied mind.[66]

Reply

Marsh does provide a strong alternative neurological explanation, but it does not entirely account for why emergent mental states should be so similar from different recovering brain states of different people at different times. Of course, for the materialist approach as taken in this book, a neurological explanation is key to explaining brain function, but emergent mental states supervene on brain states. For NDE testimonies (mental state descriptions) to be so similar between patients, the brain states must also be very similar; a state of affairs that usually results from identical or near identical perceptions. Marsh would no doubt claim that the perceptual content of an NDE is too variable to claim similarity.[67]

Memory of a NDE is sharp and stable over a long period and is frequently life-changing with the experiencer becoming more altruistic.[68] Moreover, NDEs are frequently explained by anoxia or hypoxia; the former a cessation of oxygen to the brain, the latter a reduction in the level of oxygen in the blood. Anoxia results in immediate loss of consciousness, a typical example being a heart attack. After five to ten minutes, brain damage occurs. Hypoxia is slower, but results in disorientation, confusion, and eventual unconsciousness. The clarity of testimony, heightened consciousness, and life-changing effects of NDEs are not what we would expect from a state of disorientation and confusion due to a lack of oxygen.[69]

Idiosyncratic and geo-centered descriptive variation may be the result of post-NDE embellishment as the essential features of different testimonies display remarkable similarity. Such embellishment may not be intentional, but may act to reduce dissonance caused by a highly enigmatic experience; in short, testimonies are automatically adjusted to cohere with an existing worldview. This is supported to some extent by the claim of some experiencers that their NDE is ineffable.[70]

Substance Dualism

The supernatural NDE proponents suggest that a disembodied mind floats free and travels away from the near-death patient to experience, in the first instance, events near to their body; out-of-body experience. Then this ghostly mind travels through a tunnel to a light, which is the entrance to a Heavenly realm, where they see, converse with, and touch long-dead relatives. Ignoring the problems of seeing, hearing, speaking, touching, and general orientation without a body and any physiological perceptual apparatus, this conception is that of substance dualism. Such a ghost in the machine notion is currently very unpopular with philosophers of mind,[71] although there are some philosophers who still give it support.[72] Space prevents me reviewing the arguments for and against substance dualism, but I have taken a materialist conception of mind, resurrection, and the afterlife throughout this book.

Vivid Experiences

Prima facie, the vivid descriptions of reported NDEs is persuasive, but such vividness does not imply that what is described is true; more supporting evidence is required. Notwithstanding, claims of extraordinary perceptual clarity cannot simply be dismissed as equivalent to dreams; for example, flying experiences. When we awake from such dreams we can tell the difference between the dream and reality, but this is not the case with NDEs. The reports are of perceptions that were very real to the experiencer. However, there is a differentiating factor between dreams, normal perception, and NDEs. The brain is dying and under extreme stress with NDEs and the individual may be subject to vivid hallucinatory experiences; nevertheless, the reports are large in number, generally lucid and similar in content. It seems the evidence of testimonials concerning the vividness of NDEs supports a supernatural hypothesis to some extent, but is far from conclusive.[73]

NDE and the Blind

Visual NDEs reported by blind people, even blind from birth, is most compelling.[74] The resurrection possibility-model of Chapter 7 would provide a materialistic explanation; that is, the emulated counterpart of the blind person in Heaven is not blind and as she begins to awaken she can see. As the process reverses; that is, the blind person recovers and the counterpart reverts into a non-cognitive state, the blind person, through the connectivity process, recalls the visual experience. However, even though this blind person scenario is commensurable with the possibility-model of Chapter 7, this does not ensure its actuality. Indeed, less fanciful explanations can be provided and Fischer and Mitchell-Yellin (2016) do so by distinguishing between no-input blindness and process blindness. The former is a signal input problem from the eyes and or optic nerve, the latter is neurological; namely, the inability of the brain to construct visual consciousness. Input blindness and NDEs may be explainable by visual hallucinations that can occur without a NDE. Process blindness is a different matter altogether, but there is no information correlating NEDs and process blindness.[75]

Recreational Drugs

Oliver Sacks (2012) provides a personal testimony of someone's experiences on a LSD trip. The testimony contains very similar characteristics to that of an NDE; namely, telepathy, hallucinations, lifestyle recommendations, out-of-body experience, travel through a tunnel of light, mysterious communication including a stay-or-return option, an environment of love and light, and a detailed life review.[76] This testimony does not match all of Moody's characteristics above, but there is a reasonable match. The experience was caused by a recreational drug and was not an NDE, yet the experience was very similar. Unless the LSD triggered a supernatural NDE-type episode, the hallucinatory experience was the result of a particular brain state brought on by the chemistry of a hallucinogenic drug. Although it would be unreasonable to extrapolate from one or some examples of physically induced NDE-type hallucinations to the claim that all NDE episodes are brain state induced hallucinations, it is nonetheless confirmatory evidence to this effect. However, there are some essential questions that need to be answered before any conclusions can be drawn:

1. Could LSD trigger a *supernatural* NDE-type episode?
2. Had the tripper been discussing NDEs prior to the LSD trip?
3. Had the tripper recently watched a film or read a book involving a NDE?
4. Had the tripper ever had a NDE?
5. Had a member of the tripper's family or a close friend had an NDE?
6. Was the tripper aware of the common characteristics of a NDE?

Answers to these questions are not provided by Sacks, but an affirmative answer to any one of them may provide a reason for the content of this particular LSD trip.

EVALUATORS' CONCLUSION

It is conceivable that certain perceptual-type sensations will occur in the context of a brain under stress and/or under the influence of medical drugs; for example, tunneling, repetitious sounds, bright light, and color.[77] However, perceptual detail, such as meeting dead relatives,[78] landscape features, conversations, touch, religious encounters,[79] and post-NDE personality changes [80] may be stretching the neurological explanation too far, although we do recall such things in dreams: "…it is admittedly intriguing why such a sense of reality and vividness pervades the remembered near-death event." [Marsh, 2010, 243]. Moreover, the two explanations are not necessarily mutually exclusive, particularly if a materialist explanation rather than a substance dualism explanation is adopted; for example, the materialist possibility-model of Chapter 7, where mental connectivity plays a crucial role. However, increased neural activity would be expected in such a scenario, not a cessation of neural activity as is suggested with NDEs. Interestingly, Jimo Borjigin et al (2013) has demonstrated increased activity in the dying brain, albeit in rats.[81]

The syndicate would be required to consider the likelihood ratio $\Pr(NDE|r \;\&\; k)/\Pr(NDE|\neg r \;\&\; k)$ where:

> *NDE* is the proposition: *there are perceived conscious visitations to Heaven or other supernatural realms by persons while in a near-death state.*
> *r* is the proposition: *dead persons will be resurrected*
> *k* is background knowledge associated with NDEs

If there is a Heavenly realm for persons to be resurrected into, and given the inter-world connectivity presented in the resurrection possibility-model of Chapter 7, we would expect persons on the brink of death to experience the result of a close-to-death connectivity, certainly more so than if the resurrection proposition were false. Moreover, we would also expect such experience to be similar between different NDEs; that is, not one NDE-person experiencing a heavenly environment and another person the far side of the moon. Consequently, $\Pr(NDE|r \;\&\; k) > \Pr(NDE|\neg r \;\&\; k)$. However, materialistic explanations residing in *k* are not without persuasive force, so the likelihood ratio-value would not be great.

Evaluators' Judgement

After consideration of the evidence and arguments and despite some consternation, particularly with the atheist, the evaluators believe that a consensus would be formed after assuming the truth of *r* and the truth of ¬*r* with the two likelihood functions respectively, namely that Pr(*NDE*|*r* & *k*) is greater than Pr(*NDE*|¬*r* & *k*).

Consensus

The evaluators believe that all three hypothetical members would reach a consensus as follows:

Pr(*NDE*|*r* & *k*) = 0.60
Pr(*NDE*|¬*r* & *k*) = 0.50
Pr(*NDE*|*r* & *k*)/Pr(*NDE*|¬*r* & *k*) = λ = 1.2

The likelihood values are high even though the ratio is close to unity. This is because of the persuasive force of naturalistic explanations giving Pr(*NDE*|¬*r* & *k*) considerable value. Notwithstanding, other auxiliary assumptions, for example, *dead but resurrected relatives and love ones want to make contact with the dying person*, Swinburne's principle of testimony and the similarity of NDEs provide a probabilistic edge to Pr(*NDE*|*r* & *k*) over that of Pr(*NDE*|¬*r* & *k*). Moreover, the resurrection hypothesis is a complete folk-psychic explanation of NDE evidence compared to a scientific one, albeit buttressed with auxiliary assumptions.[82] Nevertheless, the different elements of a scientific or naturalistic explanation, namely, specific physiological, psychological, and sociological elements taken in conjunction and or disjunction are also persuasive, particularly as they are commensurable with the scientific canon.[83] Readers may disagree with these individual values, but of course, it is the ratio value that is important to the Bayesian calculation.

Prior Probability

The priors from above are: Atheist Pr(*r*|*k*) = 0.52; theist (agnostic) Pr(*r*|*k*) = 0.74; theist Pr(*r*|*k*) = 0.82.

Bayesian Conditionalization

Evidence presented by the Near Death Experience Research Foundation and the Division of Perceptual Studies, University of Virginia School of Medicine,[84] suggests that NDEs do occur and that testimony shows a commonality between experiencers.[85] Given this data, the evaluators believe that all members of the hypothetical syndicate would conditionalize on the posterior function Pr(*r*|*NDE* & *k*).

Posterior Probability

From Bayes' theorem (likelihood ratio form):

$$\Pr(r|NDE \,\&\, k) = \frac{\lambda \times \Pr(r|k)}{[\lambda \times \Pr(r|k)] + [1 - \Pr(r|k)]}$$

Atheist

$$\Pr(r|NDE \,\&\, k) = \frac{1.2 \times 0.52}{[1.2 \times 0.52] + [1 - 0.52]} = 0.57$$

Theist (Agnostic)

$$\Pr(r|NDE \,\&\, k) = \frac{1.2 \times 0.74}{[1.2 \times 0.74] + [1 - 0.74]} = 0.77$$

Theist

$$\Pr(r|NDE \,\&\, k) = \frac{1.2 \times 0.82}{[1.2 \times 0.82] + [1 - 0.82]} = 0.85$$

APPARITIONS

This evidential element falls under the heading of the paranormal, a domain comprising several supernatural manifestations from ghosts to poltergeists. Such phenomena may be interrelated, but my concern is with those phenomena that may probabilistically confirm the resurrection proposition *r*. If someone 'sees' an apparition of a dead relative or a medium makes contact with a dead relative on behalf of someone then, *prima facie*, this is confirmatory evidence of human resurrection. Indeed, apparitions or ghosts and the supernatural claims of mediums would not be that surprising if human resurrection and inter-world connectivity were a reality.[86] I make this claim because, from the criteria of diachronic person-identity of Chapter 6 and the resurrection possibility-model proposed in Chapter 7, resurrected persons would be aware of the world they once inhabited. They would remember it, particularly their former homes, friends, and relatives and if able, would make contact, just as we do in our world. I have argued for such inter-world connectivity in terms of a divine/human connectivity, but perhaps this can be extended to include person-to-person inter-world connectivity. The possibility-model of human resurrection adopted in Chapter 7 is based upon such connectivity and a supernatural view of near-death experiences is also commensurable with this view.

Given the materialistic view I have adopted in this book, paranormal activity cannot be explained by a disembodied spirit-based notion, but rather by a materialist inter-world connectivity, whereby apparitions, for example, are person-to-person caused hallucinations that may occur with several experiencers simultaneously.[87] This clearly does not seem to explain all supernatural claims; for example, physical objects suddenly moving and poltergeist hauntings, which *prima facie*, do not seem to be hallucinatory. However, if such paranormal events do occur, they may be related in some way to this materialistic inter-world connectivity. Alternatively, it may be the result of psycho-kinesis from living persons.[88] In fact, hallucinations may be one of several inter-world effects. Raynor Johnson (1967) suggests there are both objective and subjective apparitions, the former being subject to third-party verification such as photography.[89] This highlights the notion of veridical or verifiable apparitions which provide explanatory difficulties for the sceptic. Unknown information that can be verified post-sighting fall into this category; for example, unrecognized features, movements, clothing, and smell of the deceased person. Certainly, photographs and multiple witness sightings are examples of veridical apparitions.

The inter-world connectivity in the possibility-model of Chapter 7 could account for some of these veridical apparitions, but photographs of apparitions and poltergeists are more difficult. Perhaps there is a way that beings in this other parallel world can cross the threshold to our world. Although speculatively interesting, this diverges from the theme of human resurrection.

The notion of a hallucination conjures a view of an abnormal brain-state giving rise to a false perception. Oxygen deprivation, hallucinatory drugs, alcoholism, psychosis, brain damage, and often just drowsiness are associated with the occurrence of a hallucination.[90] Moreover, animism and anthropomorphism as instinctive perceptual responses to the natural world may account for false perceptions, as for example, seeing human faces or animals in cloud formations.[91] These are cases of an illusion rather than a hallucination, as a physical object or shadow is the primary cause of the false perception. Hallucinations are not caused by the presence of an object or an environmental change, but by the action of the brain.[92] The notion of a hallucination also applies to paranormal apparitions such as seeing a dead relative, being touched, sensing a strange smell, or hearing a voice. The dead relative is not physically present in front of the experiencer, rather the experiencer falsely perceives the dead relative due to a causal influence on her brain; her brain-state changes and a perceptual mental-state emerges. This is no different to other extrinsic causal factors such as drugs, but in this case, it is a paranormal cause; such a paranormal connectivity cause has been described as telepathic.[93] Thus, hallucinations, which should be differentiated from dreams,[94] illusions, and pseudo-hallucinations,[95] are false perceptions that occur concurrently with and possess the same qualities as true perceptions; that is, they are vivid, substantial, with apparent spatio-temporal location.[96]

Interestingly, many reported apparitions are not of deceased persons, but of persons living in this world;[97] it seems that mental connectivity (or telepathy) is not limited to inter-world connectivity. Thus, assuming the veracity of a materialistic human resurrection hypothesis as presented in this book, I define a paranormal apparition as:

> *An hallucination (manifested in a variety modes) experienced by one or more persons and caused by mental connectivity with other living persons, either in this world or resurrected in another.*

There are alternative paranormal hallucination explanations other than being caused by a resurrected person such as a naturalistic recording, whereby an incident is held in the environment causing the percipient to hallucinate as a form of playback.[98] Demonic explanations are also a variant; the apparitions are caused by non-human supernatural malevolent beings.[99] The former explanation is not commensurable with the above definition of a paranormal apparition, the latter is, but both are beyond the scope of this short introduction.

Determining the veracity of paranormal claims takes several forms, such as personal knowledge from an unexpected spontaneous apparition, which is a subjective assessment. Planned ghost-hunts, sometimes involving a vigil by a group of observers with recording equipment, adds more objectivity to the process. Individual case investigations also can add objectivity to individual testimony. Testimony-based surveys are a way of objectifying individual experience and many substantial surveys have been undertaken (see Bayesian Conditionalization below).

> Typically, the survey population is asked to respond to questions such as: "Have you ever, when believing yourself to be completely awake, had a vivid impression of seeing or being touched by a living being or inanimate object, or of hearing a voice; which impression, so far as you could discover, was not due to any external physical cause?" [Sidgwick & Johnson et al, 1894]

Counter Argument

The first counter argument for the use of paranormal apparitions as evidence in a Bayesian analysis of the resurrection proposition r is that such apparitions are uncanny experiences. As such they fall within the religious experience domain introduced in Chapter 5, where such experiences are used as evidence in support of the theistic proposition *God exists*. The *uncanny experience* domain includes many relevant experiential occurrences such as divine connectivity, but also includes seeing a ghost. If uncanny experiences are used to confirm probabilistically the theistic proposition p, and p is then used to confirm the human resurrection proposition r, then uncanny experience should not be used again to probabilistically confirm r; it is a case of double counting.

Reply

Paranormal apparitions are uncanny experiences, but they have little individual confirming or disconfirming force on the theistic proposition p. $\Pr(a|p \,\&\, k) \approx \Pr(a|\neg p \,\&\, k)$; that is, we would just as likely find evidence of a (apparitions) whether God exists or not. This is liable to be the case for all uncanny experiences, save those that result in a dramatic conversion experience DCE;[100] ghost sightings are most unlikely to result in a DCE. I would argue that paranormal apparitions as evidence for r are so removed from the confirming force of DCEs on p, that any accusation of double counting is unfounded.

Counter Argument

The claim that ghosts are hallucinations indicates that the experiencers are suffering from some physical or psychiatric disorder rather than experiencing an objective reality; it is a case of delirium not the paranormal.[101] Even a less dramatic causal explanation such as depression, anxiety, drowsiness, or excessive alcohol is a better inference to the best explanation than the paranormal. Indeed, the "...commonest examples of hallucinations are hypnagogic and hypnapompic hallucinations, which occur [as a state of consciousness] at the onset of sleep or on waking respectively. They are usually normal phenomena..." [Chiu, 1989, 293] Furthermore, our instinctive tendency to see animal or human forms in innocuous objects, such as the tree stump seen as a wolf, would probably account for some reported sightings. This may account for why reports of ghosts are traditionally associated with antiquated buildings, dark woods, gloomy back streets, and spooky basements; that is, suggestibility coupled with anthropomorphic illusions.[102]

Reply

Although hallucinations do result from drowsiness, excessive stress, and mental and physical disorders, they can and do occur in the absence of these extrinsic natural causes. The Cardiff Anomalous Perception Scale was employed by Vaughan Bell et al (2011) to determine the distress, intrusiveness, and frequency of perceptual distortions with clinical and non-clinical populations. Significantly, one of their findings confirmed "that anomalous perceptual experiences are common in the non-clinical population and that a significant minority have greater levels of perceptual distortion than patients with psychotic diagnoses." [Bell et al, 2011, 455]

Testimony of paranormal apparitions have been made where more than one experiencer 'saw' the apparition which reduces the probability of the cause being due to a mental or physical disorder,[103] and many surveys have eliminated reported cases where the experiencer was liable to be in a drowsy state such as being in bed. Also, there are numerous reports of repeated sightings of the same apparition by different witnesses at different times (hauntings), again reducing the probability of the anthropomorphizing of natural objects.[104] Anthropomorphizing is illusionary rather than hallucinatory, as it is a false perception of an object or shadow. I believe that paranormal apparitions are primarily hallucinations through intrinsic brain-state change. Notwithstanding, an anthropomorphizing illusion could explain some 'sightings.'

The suggestibility argument is not supported by the evidence. In the Romer & Smith (2009) survey, 70.5% were at-home reports, and 29.5% not-at-home reports with approximately 25% of those in the car.[105]

Naturalistic explanations are persuasive and Richard Wiseman (2015) provides a clear explication of the many possible naturalistic causes for false paranormal experiences. Naturalistic explanations of why the mind hallucinates or is subject to an illusion are not surprising; the brain is subject to multiple causal effects that can result in similar emergent mental-states. What is surprising is the claim that only one cause is exhaustive. The mental-state of the taste of sweetness supervenes on a brain-state caused by sugar on the tongue. However, the same mental-state can be caused by saccharine on the tongue; it would be bizarre to claim that the only possible cause of the sensation of sweetness is sugar. Seeing a ghostly figure can be caused by anthropomorphizing a shadow, an unpleasant drug experience, self-delusion after a scary film, excessive alcohol, drowsiness, suggestibility on a ghost-hunt, or an actual paranormal hallucination. The brain-state is similar in all cases as is the emergent mental-state. Selection of the probable cause must be based on inference to the best explanation founded upon the evidence available.

EVALUATORS' CONCLUSION

Subject to evidence, there is a case for paranormal apparitions to be considered as confirming evidence for the human resurrection hypothesis *r*.

However, this requires "…additional assumptions about survivors [the resurrected] having the interest, intention, and power to communicate with the living, their retaining substantial knowledge of their ante-mortem existence, and their having a certain degree of epistemic access to the minds of living persons and states of affairs in the empirical world that occur after their death." [Sudduth, 2016, 199]

Diachronic person-identity explored in Chapter 6 revealed the importance of both psychological and physical continuity in a sound diachronic person-identity model. Further, the resurrection possibility-model presented in Chapter 7 preserved both the psychological and embodiment requirements of diachronic person-identity.[106] Indeed, "If we conceive of survival [resurrection] in purely physical terms, as embodied existence, the issues of causal agency and perception do not present any initial difficulty." [Sudduth, 2016, 37] Thus, if there is a desire to communicate with friends and loved ones in this world, then from the arguments of Chapters 6 and 7, there is no reason to think that the same motives will not continue post-mortem; an interactionist hypothesis. This able-and-willing assumption does not imply that all resurrected persons would want to communicate; radio-silence is not uncommon in our world. However, the ability to communicate across worlds is another question and that, of course, relates to the evidential element under consideration.

Michael Sudduth (2016) argues that "What is required is evidence of what persons would be like if they should survive death, where this evidence is independent of the assumption that persons survive death." [*Ibid*, 304] However, I would argue that reasoned argument provides equal independent justification and my arguments of Chapters 6 and 7 provide good reason for us to adopt an interactionist hypothesis.

The evidence of the Resurrection (of Jesus) is an example of post-mortem communication and the naturalistic possibilities of cryonics, digital resurrection, and androids would support post-mortem communication if not cross-world communication.

In light of the above, the hypothetical syndicate, through the judgement of the evaluators, must consider the likelihood ratio $\Pr(a|r \& k)/\Pr(a|\neg r \& k)$ where a is the evidential proposition *living persons experience apparitions of dead persons*.

Evaluators' Judgement

The evaluators believe that the syndicate would accept, subject to evidence, that $\Pr(a|r \& k) > \Pr(a|\neg r \& k)$. However, there is a probability that a will be true even if r is false; for example, demonic causation and naturalistic causation, such as fraud and coincidence. Consequently, the evaluators consider that the syndicate would not accept a large discrepancy between the two functions.

Consensus

The evaluators believe that all three hypothetical members would reach a consensus as follows:

$\Pr(a|r \,\&\, k) = 0.85$
$\Pr(a|\neg r \,\&\, k) = 0.65$
$\Pr(a|r \,\&\, k)/\Pr(a|\neg r \,\&\, k) = \lambda = 1.31$

Prior Probability

The priors for $\Pr(r|k)$ from above are: atheist = 0.57; theist (agnostic) $\Pr(r|k)$ = 0.77; theist $\Pr(r|k) = 0.85$.

Bayesian Conditionalization

I list below the basic results of the main surveys concerned with paranormal apparitions.

Table 8.4. Paranormal Surveys

Survey (See Bibliography)	*Country*	*Survey Type*	*Sample Size*	*% +ve Responses*
Sidwick et al (1894)	UK	Questionnaire	17,000	9.9
West D. J. (1948)	UK	Questionnaire	1,519	14.3
Palmer J. & Dennis M. (1974)	US	Postal	1,000	17
Haraldsson E. et al (1977)	Iceland	Postal	902	31
Blackmore S. J. (1984)	UK	Postal	595	45
West D. J. (1990)	UK	Questionnaire	850	11.3
Tien A. Y. (1991)	US	Interviews	18,572	13
West D. J. (1995)	UK	Telephone	1,002	22
Ipsos MORI (1998)	UK	Telephone	721	15
Ipsos MORI (2007)	UK	Telephone	1,105	13.8
YouGov (2011)	UK	Telephone	2,040	25

This table summarizes the positive responses from several surveys with different sample sizes to a typical apparition question, such as "have you seen an apparition?." The final column in the table shows the percentage positive responses for each survey. Reference to each survey can be found in the Bibliography.

It can be seen from the above table that a sizable number of apparitions have occurred for the populations captured by the different surveys. A considerable number were reported apparitions of dead persons. This evidence does not entail that such apparitions are of dead people, just that they are believed to be so by those experiencing them. From this, the evaluators conclude that the syndicate will Bayesian conditionalize on their resultant posterior probability values.

Posterior Probability

From Bayes' theorem (likelihood ratio form):

$$\Pr(r|\mathrm{a}\ \&\ k) = \frac{\lambda \times \Pr(r|k)}{[\lambda \times \Pr(r|k)] + [1 - \Pr(r|k)]}$$

Substituting the three prior probabilities of the hypothetical syndicate members into the theorem:

Atheist

$$\Pr(r|a\ \&\ k) = \frac{1.31 \times 0.57}{[1.31 \times 0.57] + [1 - 0.57]} = 0.64$$

Theist (Agnostic)

$$\Pr(r|a\ \&\ k) = \frac{1.31 \times 0.77}{[1.31 \times 0.77] + [1 - 0.77]} = 0.81$$

Theist

$$\Pr(r|a\ \&\ k) = \frac{1.31 \times 0.85}{[1.31 \times 0.85] + [1 - 0.85]} = 0.88$$

CONCLUSION

I centered the human resurrection hypothesis on the existence of God and as such, the theistic proposition p represented the starting position for this chapter. The probability of *p*, being established in Chapters 3, 4, and 5, was first employed to establish the conditional probability of the Resurrection (of Jesus) proposition *R* given *p* and background knowledge. Then the posterior probability of the resurrection proposition *r* given *p*, *R*, and background knowledge was determined. Of additional interest was the probability of a naturalistically caused resurrection scenario.

The next evidential element considered was near-death experience (NDE) and it was concluded that all members of the hypothetical syndicate would accept that NDE, given similarity in testimony of NDE patients, would have some confirming effect on *r*.

The final evidential element was that of paranormal apparitions, and the evaluators believed that all three members of the syndicate would form a consensus such that this evidence confirmed *r*.

The final result of the probabilistic analysis for the hypothetical theist (agnostic) and theist is that human resurrection is highly probable and for the atheist much more probable than not. This surprising result for the atheist is mainly due to the naturalistic possibilities for person-resurrection, although these are futuristic possibilities. However, the 2012 survey conducted by Alice Sullivan, David Voas, and Matt Brown reveals the reality of a belief in human resurrection without belief in God.[107] Of further interest is that the theist (agnostic) and theist converged in their degree of belief in human resurrection; moreover, their degree of belief is very high.

NOTES

1. By inter-probabilistic, I mean the probability value of a proposition determined, in part, by the probability value of a different but related proposition.
2. Paterson R. W. K. (1995); pp. 176–185 and Ducasse C. J. (1961); Ch. XXI.
3. A Bayesian network is an interrelated graphical model comprising different states of affairs expressed by propositions and the network describes how those propositions are interrelated by probabilities. Bayesian networking is associated with Judea Pearl (1985).
4. Swinburne R. (2003[A]); Ch. 2.
5. Miracles; Sec. 2; Ch. 5. 'Highly extreme…' compared to 'extreme…,' e.g., walking on water or turning water into wine, because the Resurrection is, at an enigmatic level, above such events.
6. Swinburne R. (2003[A]); p. 190.
7. More precisely, incarnation as prescribed by the Council of Chalcedon: Swinburne R. (2003[A]); Ch. 2.
8. *Ibid*; p. 211.
9. The Logical Theory of Probability; Ch. 2.
10. A Further Bayesian Analysis; Ch. 4.
11. Ducasse C. (1961); p. 193; verbal instead of numerical.

12. The super-miracle being the death and Resurrection of Jesus: Swinburne R. (2003A); p. 211f.
13. *Ibid*; p. 212.
14. This does assume exotic naturalistic auxiliary assumptions of a 'science fiction' genre are unjustifiable. See Appendix 1: The Auxiliary Assumption Requirement.
15. Paterson R. W. K. (1995); p. 160; although no numerical prior probability-value is quoted.
16. I say futuristic, but Steinhart reflects upon the possibility that we are already in a digital simulation: Steinhart E. (2014); Ch. V.
17. *Ibid*; pp. 57–59.
18. *Ibid*; pp. 34–39; p. 43.
19. *Teleportation;* Ch. 7.
20. Steinhart E. (2014); pp. 23; 45; 48–49; 64.
21. *Ibid*, p. 64f.
22. *Ibid*, p. 28.
23. *Ibid*, p. 65.
24. Ch. 6; Identity.
25. Chs. 6 and 7; *The Only x and y Principle.*
26. Ch. 7; Genuine Possibility; *Nomological possibility*; *Emulation.*
27. Ch. 1; Omni-be nevolence; *Suffering and Salvation* and Steinhart E. (2014); p. 72f.
28. Steinhart E. (2014); p. 87.
29. *Ibid*; pp. 94–105; compare this to the landscape universe: Ch. 7; *Heaven.*
30. That is, with a boundary; a causal boundary (c-boundary) where cause and effect ceases, i.e., the edge of space-time. Tipler's assumption is that the universe will contract to a singularity, the Omega Point; a single point c-boundary.
31. Greene B. (2011); Ch. 10 for a similar simulated multiverse hypothesis.
32. Tippler F. J. (1994.
33. It is not that the galaxies are themselves accelerating away through space, but that the expansion of space is accelerating.
34. Constitutionalism; Ch. 7.
35. The First-Person Perspective; Ch. 6.
36. I am using different symbols than Meierding for consistency.
37. Over-determination expresses an excess of causal contributors to an effect whereby any one of the contributors could cause the effect.
38. This refers to Mark 16:9–20 added later by an unknown copyist.
39. Paul's reference to 'the twelve' in 1 Corinthians 15 included Matthias who had replaced Judas.
40. This appearance may have been apparitional as opposed to a bodily appearance.
41. Swinburne R. (2003A); p. 156ff.
42. *Ibid*; p. 159.
43. Habermas G. R. & Licona M. R. (2004); pp. 69–74.
44. Swinburne R. (2003A); pp. 57–59.
45. Matthew 28; 11–15.

46. Habermas G. R. & Licona M. R. (2004); p. 262n.

47. Compare Mark 16: 1–8 with Peter 39–42.

48. Habermas G. R. & Licona M. R. (2004); p. 52f.

49. Not all reported NDEs are close to death experiences; Owens J., Cook E. W. and Stevenson I. (1990).

50. Of particular note, because of his profession, is that of Dr. Eben Alexander (2012).

51. Moody R. A. (2001); Ch. 2.

52. Lommel van P. (2010); pp. 19–26.

53. Moody R. A. (2001); pp. 29–39.

54. *Ibid*, pp. 116–122.

55. Satori P. (2014); p. 39–45.

56. Morse M. and Perry P. (1990); p. 114f.

57. Burpo T. with Vincent L. (2010).

58. Fischer J. and Mitchell-Yellin B. (2016); Ch. 11.

59. The 2010 book *The Boy Who Came Back from Heaven* is a case of intentional deception.

60. Marsh M. N. (2010); p. 73.

61. *Ibid*; p. 74.

62. *Ibid*; p. 76f.

63. Fischer J. & Mitchell-Yellin B. (2016); Ch. 3.

64. Marsh M. N. (2010); Chs. 5–9.

65. *Ibid*; pp. 88–91.

66. *Ibid*; pp. 92–94.

67. *Ibid*; pp. 92–94.

68. Satori P. (2014); pp. 30–34.

69. *Ibid*; pp. 111–13.

70. Moody R. A. (2001); p. 15f.

71. Maslin K. T. (2007); Ch. 2.

72. Foster J. (1996).

73. Fischer J. & Mitchell-Yellin B. (2016); Ch. 5.

74. Long J. with Perry P. (21010); Ch. 5.

75. Fischer J. & Mitchell-Yellin B. (2016); Ch. 6.

76. Sacks O. (2012); p. 101f.

77. Marsh M. N. (2010); Ch. 9.

78. The experience of meeting dead relatives and loved ones during a NDE could be explained by the need for comfort in highly stressful circumstances; circumstances of terror: Fischer J. & Mitchell-Yellin B. (2016); pp. 68–73.

79. Alexander E. (2012); pp. 38–41; 45–49.

80. Lommel van P. (2010); Ch. 3.

81. Borjigin J., Lee U.; Liu T., Pal D., Huff S., Klarr D., Sloboda J., Hernandez J., Wang M. M. and Mashour G. A. (2013).

82. Note that I avoid appeal to the principle of simplicity, which I argue against as an *a priori* principle in Chapter 3; *Simplicity*.

83. Augustine K. (2015); Ch. 25 for a detailed hallucinatory explanation of NDEs.

84. www.NDERF.org and https://med.virginia.edu/perceptual-studies/research-area/near-death-experiences-ndes/.

85. http://www.nderf.org/NDERF/Research/Research_Overview_Right.htm for a detailed list of NDE research. Also Long J. with Perry P. (2010).

86. I am assuming medium supernatural connectivity is a paranormal living to dead connectivity like NDEs.

87. Romer C. J. (2011); pp. 100–106.

88. Johnson R. C. (1989); pp. 248–251.

89. Johnson R. C. (1967); p. 102f.

90. Chiu L. P. W. (1989).

91. Guthrie S. (1993).

92. Johnson R. C. (1967); p. 207.

93. Tyrrell G. N. M. (1973); p. 42f; Evans H. (2002); p. 35f; Johnson R. C. (1967); Ch. 6; Almeder R. (1992); pp. 112–117.

94. Puhle A. and Parker A. (2017); pp. 145–160 for a study of paranormal dream experiences.

95. Pseudo-hallucination lacks the vividness present in normal perception Chiu L. P. W. (1989); p. 293.

96. *Loc. Cit.*

97. Smith R. (2013); p. 267f.

98. Myers, F.W.H. (1903)—*Human Personality and its Survival of Bodily Death*—London: Longmans Green.

99. For these alternative explanations see McCue P. (2002)—*Theories of Haunting: A Critical Overview*—in *Journal of the Society for Psychical Research* 6.

100. Dramatic Conversion; Ch. 5.

101. Chiu L. P. W. (1989); p. 293ff.

102. Wiseman R. (2015); pp. 226–228.

103. The 1894 Census of Hallucinations recorded 95 multiple witness hallucinations.

104. Tyrrell G. N. M. (1973); pp. 69–73.

105. Smith R. (2013); pp. 139–141; Romer C. J. (2011); p. 116f.

106. Sudduth M. (2016); pp. 36–46 explores various interactionist survival models; that is, cross-world connectivity scenarios between the resurrected and the living (Earth-bound) comprising strong, moderate, and weak psychological survival profiles.

107. Sullivan A., Voas D. & Brown M. (2012); p. 17f.

Conclusion

Stewards' Inquiry

The vicarious Bayesian analysis of the resurrection hypothesis has resulted in three probabilistic values for the resurrection proposition *r*; *dead persons will be resurrected*. Two of these values are both high and close in value; namely, the theist (agnostic) and the theist; the atheist's value is also quite high. These results bear a caveat, in so far as they are degrees of belief in *r* rather than an objective probability value. This does not entail that such subjective probability values are value-less, on the contrary, adherence to the axioms of probability enabled objectivity to enter the process of forming these beliefs based on evidence.

Adherence to the axioms of probability was not the only objectifying process in the analysis; the application of vicarious probability (an off-shoot of intersubjective probability), governed by evaluators with opposing theistic convictions, added further objectivity to the analysis. Moreover, the cumulative Bayesian process ensured a coherent and progressive chain of analysis through a wide evidential set. This included seven evidential elements for and against the existence of God, the resurrection of Jesus as possible evidence of human resurrection, near-death experiences, and apparitions (see Appendix 2). Further, the evaluators gave due consideration to a range of arguments associated with all the evidential elements; for example, those arguments concerning the problem of evil and the cosmological argument and five others probabilistically related to the theistic proposition *God exists*. Likewise, associated arguments related to the resurrection proposition *dead persons will be resurrected*. Additionally, when it was thought by the evaluators that consensus would not be reached, they defaulted to subjective probability for those hypothetical syndicate members who would be unlikely to concur with the other members. Thus, consensus was not enforced by the evaluators when it was clearly inappropriate to do so.

The description, application, and justification of auxiliary assumptions (see Appendix 1) was also included, where such assumptions were independent of the evidence under consideration. Their justification was, overall, based upon reasoned argument and embedded in the various arguments presented. Such auxiliary assumptions were deemed to reside in the background evidence *k* in the probability functions employed.

Given the above, it is my conclusion that, for those who hold a theistic or near-theistic conviction, belief in human resurrection is warranted. For those who hold an atheistic conviction, belief in human resurrection might be surprisingly high and not to be rejected out-of-hand. Perhaps for the atheist, a more naturalistic, rather than theistic possibility for human resurrection may be more acceptable; for example, Tippler's Omega Point Theory, cryonics, digital afterlives, synthetic emulation through androids, or a Buddhist-type spiritual resurrection hypothesis that does not involve the will of, or even the existence of God.

The theistic resurrection hypothesis presented is a materialist one; that is, there is a *bodily* resurrection in Heaven. Taking such a position involves quite a complex possibility-model to permit the hypothesis to adhere as close as possible to the requirements of diachronic person-identity presented in Chapter 6. A dualistic hypothesis of body and soul would be far simpler and, like others, I find such a view difficult to adopt,[1] however I do not rule it out entirely (1 Corinthians 15: 42–49).

The argument is one of individual resurrection upon death, but the hypothesis could also account for a general resurrection in an eschatological doctrine; that is, the resurrection of all mankind at the end of time. The divine emulation process proposed in the possibility-model of Chapter 7 could involve a delay in the awakening of the emulated person for a general resurrection scenario or, a repeat of the process in a dual personal and general resurrection scenario.

The issue of animal resurrection has not been addressed, being beyond the scope of this book, but suffice it to say, there is no reason, given the resurrection possibility-model of Chapter 7, that animal resurrection would not happen.

A further issue I have not addressed is that of quality assurance in human resurrection. By this I mean that the whole brain emulation process introduced in Chapters 6 and 7 will capture the personality of the emulatee. If that personality is offensive or despicable, then Heaven would possibly be welcoming an unwanted guest. There are two possibilities for ameliorating this unfortunate scenario; first, the Earth-bound person who has a strong belief in the resurrection hypothesis can help ensure that their own character is suitable for a continuing life in Heaven; a notion that echoes John Hick's 1981 concept of *soul-making*. In the case of Christianity, the quality assurance requirements are clearly laid-out in the Gospels in the reported life and teachings of Jesus; for example, treating others as you would want to be treated yourself. For the Christian, Jesus is a paradigm of what is ideal for the character of a potential Heavenly resident and the individual can take some responsibility for her own pre-resurrection development from this divine template. Other religions have their own quality assurance templates that the ante-mortem individual can adapt to. Second, the post-mortem resurrected person with an unsuitable character could be provided with further opportunity for personal development; an omni-benevolent deity would hardly write-off the unsuitable, particularly when poor character is the result of extrinsic factors. The Catholic concept of Purgatory is commensurable with such a notion.

The cumulative Bayesian analysis that culminated in the vicarious probability values for the resurrection proposition commenced with a set of prior probabilities for the theistic proposition *God exists* for a hypothetical atheist, agnostic, and theist. Indeed, most people possess such a prior degree of belief before any reflective consideration of evidence. They may have a strong religious faith or disbelieve that God exists or are undecided. If these positive and negative beliefs are formed before consideration of the relevant evidence, what justifies them and how are they formed?

Like most beliefs, they become established through the socialization process of living in a culture, being brought-up in a certain way, being formally educated from early childhood, mixing with friends, and engaging in a particular lifestyle and many other activities and relationships. The entire belief-forming process begins from early childhood, developing into a congruous nexus of inter-related beliefs; a world-view or background for interpretation. New beliefs that are congruous with this nexus are easily absorbed and those that are not too incongruous absorbed with adjustment of the belief-nexus so that the new belief dovetails into the nexus. This is not the case with strong incongruous ones; they cause a sense of dissonant discomfort and are frequently rejected out of hand. Think of the dismissive smirk of an atheist when the question of God's existence arises and the unctuous expression of the theist when their faith is questioned by a disbeliever. Such rejection is not limited to the world of religion; it happens in science, the arts, and all belief-laden pursuits.

Positive and negative religious beliefs are usually formed in the same way as many other fundamental beliefs; that is, not from formal or theoretical consideration of the relevant evidence, but through engaging in a lifestyle from early childhood and then throughout life: praxis. An adolescent raised in a religious culture may be a believer at that time, but new friends, education, and new activities may slowly change her world-view so that religious beliefs find no berth in her new adult belief-nexus. Of course, the reverse can be true, and once this belief-forming process is understood, an atmosphere of détente between the atheist and theist may emerge; hence, the varied conviction approach I have taken throughout.

This is not to claim that theoretical consideration holds no truck; it does, and a belief-nexus can be reinforced or made malleable to new ideas through such formal considerations. Indeed, the hypothetical agnostic in this study adapted their belief-nexus from agnosticism to theism through the formal consideration of the relevant evidence. Of course, the agnostic did not actually exist and so no actual belief change occurred, but the vicarious process employed demonstrated that it could easily happen.

Given the above, prior degrees of belief in an idea before any formal consideration of relevant evidence is normal and warranted, provided one's mind is not closed to either the confirming or disconfirming force of the evidence on the truth of that belief. Notwithstanding, such theoretical considerations are, by no means, the whole story, "Deciding on whether to accept a theistic [or atheistic] worldview cannot be a purely intellectual decision made at a safe distance from the context of life and action." [Cottingham J. (2015); p. 61]

Although an atheist's belief in human resurrection is more liable to be naturalistic and futuristic, I feel confident in assuring the theistic reader that they have every reason to believe that human resurrection is not only a possibility, but is also highly likely.

NOTE

1. Martin M. and Augustine K. (2015) (Eds); Part 1; Chs. 1–11.

Appendix 1

The Auxiliary Assumption Requirement

The reader will note that I have raised the issue of auxiliary assumptions in my analysis and I should like to say a little more about them. The auxiliary assumption requirement is associated with the Duhem–Quine Thesis;[1] that is, it is impossible to test a scientific hypothesis in isolation; it requires several background assumptions termed auxiliary assumptions or bundles of hypotheses.[2] Applying this notion to human resurrection, Michael Sudduth (2016) argues that all likelihoods require auxiliary assumptions to be well-defined; that is, a clear and justifiable probabilistic relationship between hypothesis and evidence.[3] He argues that:

> "…a simple hypothesis of personal survival [human resurrection], one that just postulates a surviving individual soul, person, mind, or consciousness, generates no well-defined likelihood, because it has no consequences for which we should expect to find in the way of observable evidence or features of the empirical world." [Sudduth, 2016, 215]

Thus, on its own, a resurrection hypothesis does not lead us to expect the evidence often cited; additional auxiliary assumptions are required to generate such expectation.[4] For example, testimonial evidence of apparitions would seem to have no evidential credence based on a resurrection hypothesis without suitable auxiliary assumptions to buttress the probabilistic relationship; for example, but not exclusively, a supernatural afterlife. In my argument, I refer to testimonies of near-death experiences (NDEs), apparitions, and scriptural testimony of the Resurrection of Jesus. One auxiliary assumption related to the evidence of testimony is Richard Swinburne's principle of testimony, that is, "…(in the absence of special considerations) the experiences of others are (probably) as they report them." [Swinburne, 2004, 322].

Most resurrection arguments assume that resurrected persons are spirit-like substances. In the relevant likelihood, a conjunction of a resurrection hypothesis and spirit-like auxiliary assumption without further qualifying assumptions such as a spirit/material causal explanation, would not lead us to expect the observable evidence claimed.[5] The resurrection hypothesis herein does not include a spirit-ontology, nevertheless other qualifying auxiliary assumptions are required; for example, across-world connectivity.

GENUINE POSSIBILITY

There must be a range of what counts as a well-defined likelihood from not at all to very well-defined and it seems to me the position on this continuum is subjective. Perhaps if auxiliary assumptions create a genuine possibility of a probabilistic relationship between the hypothesis and the evidence, then the likelihood can be said to be well-defined and the level of being well-defined left to a subjective or intersubjective view. I return to Scott Sturgeon's (2000) criteria for genuine possibility;[6] that is, that the probabilistic relationship between the hypothesis and the evidence in a likelihood function is well-defined if it is not fractured for logical, conceptual, metaphysical, and nomological reasons. Thus, on the grounds of nomological possibility I can see why Sudduth believes that a spirit-like conception of human resurrection does not create a well-defined probabilistic relationship between a resurrection hypothesis and his evidential set under consideration.[7]

For the evidential elements presented in my argument, I believe I have presented sufficient reasons for there to be a genuine possibility for there to be a probabilistic relationship between the evidential elements presented and their relevant confirming or disconfirming hypotheses. Thus, I argue that the likelihoods in question are well-defined.

Auxiliary assumptions need not be overtly included in the likelihood function, such as $\Pr(e|p \,\&\, a_1 \ldots \& \ldots a_n \,\&\, k)$, but can reside in background knowledge k. Notwithstanding, a good argument for human resurrection should present, in detail, these auxiliary assumptions.

The application of Sturgeon's criteria is a useful way of assessing whether a likelihood is well-defined or not. By way of an example, consider the following:

h: there is a large telescope housed in Greenwich
e: Mars follows an eccentric orbit around the sun

The formal likelihood function is $\Pr(e|h \,\&\, k)$ and a subjective likelihood based upon *h*, *e*, and *k* can be stated propositionally as follows:

p: Given that there is a large telescope housed in Greenwich, then it is probable (or improbable) that Mars follows an eccentric orbit around the sun.

If the hypothesis *h* confirms or disconfirms the evidence *e*, then it is because I have come to believe or disbelieve *e* to some degree because of *h*. However, the existence of a large telescope housed in Greenwich is irrelevant to whether Mars follows an eccentric orbit around the sun. *p* expresses an ill-defined likelihood function, but does it fulfill Sturgeon's criteria?

There is no logical contradiction in *p* and the meanings of the concepts employed do not contradict each other. The metaphysical criterion precludes genuine possibility if the essential properties of the concepts employed in the proposition are incongruous, but they are not and *p* passes the metaphysical criterion. However, the nomological criterion seems questionable as *p* suggests that a causal relationship exists between the existence of the telescope and the eccentric path of Mars. Consequently, it is reasonable to claim that the likelihood function is ill-defined.

There are auxiliary assumptions that can be added to the likelihood that remove this causal allusion, changing its ill-defined nature to a well-defined one; consider the following auxiliary assumptions:

a_1: over a period, the telescope was made to follow the path of Mars.
a_2: a proficient astronomer recorded the path of Mars around the sun.
a_3: the astronomer published her findings in a learned journal for astronomy.

The likelihood function $\Pr(e|h \ \& \ a_1 \ \& \ a_2 \ \& \ a_3 \ \& \ k)$ now fulfills all of Sturgeon's criteria and becomes well-defined. The conjunction of propositions a_1 & a_2 & a_3 can be included in *k*.

The essential auxiliary assumptions in this example are empirical; namely observational, but analytical ones would suffice; for example:

A_1: The observatory housing the telescope was also the home of a mathematician.
A_2: The mathematician theoretically determined the orbit of Mars.

Analysis, reason, and argument can also establish an essential auxiliary assumption to make likelihoods well-defined; even better, a combination of both.

In addition to the likelihood of exhibiting genuine possibility through adopted auxiliary assumptions, the adopted auxiliary assumptions have themselves to be justifiable. For example, consider the following auxiliary assumption:

a_4: Martians secretly built a telescope in Greenwich to celebrate celestial travel of an eccentric nature.

a_4 would be an auxiliary assumption that would, *prima facie*, make the likelihood well-defined. However, such an assumption begs belief and would neither be empirically nor analytically verifiable. This demonstrates that adopted auxiliary assumptions exhibit a range of justifiable forms from totally unjustified to totally justified, the Martian example being of the former category. Herein lies the danger of bulking-up the likelihood with unjustifiable auxiliary assumptions to leverage the probability in one's favor.[8] This danger also looms with the argument for human resurrection, but is mitigated by means of intersubjective probability through washing-out irrational assumptions through discussion and consensus; nevertheless, justification of auxiliary assumptions will be subjectively variable depending on prior beliefs. Notwithstanding, "What is needed is not the invention of auxiliary propositions…but the identification of auxiliary information that is independently supported." [Sober E. (2008); p. 168] This notion of independent support is crucial to any resurrection hypothesis and it can be both empirical and/or analytical.

AUXILIARY ASSUMPTIONS AND THEIR JUSTIFICATION

Throughout my cumulative Bayesian argument, I have employed likelihood ratios and, *prima facie*, some of the likelihoods appear ill-defined. However, in the background knowledge *k*, I have assumed auxiliary assumptions presented in the arguments prior to the likelihood ratio assessments and these were accounted for by the two evaluators. The essential auxiliary assumptions embedded in my overall argument that are supportive of the resurrection hypothesis are as follows:

1. Human resurrection is a genuine possibility: *see Chapters 6 & 7*
2. Resurrected persons are corporeal, not spirit-like: *see Chapters 6 and 7*
3. Resurrected persons maintain cross-world identity with a mental capacity equivalent to the peak of their ante-mortem state: *see Chapter 7*
4. Some resurrected persons can communicate to some degree with living persons: *see Chapter 7; Inter-world Connectivity and Chapter 8: Apparitions*
5. Some resurrected persons possess the desire and intention to communicate, to some degree, with living persons: *see Chapter 8: Near Death Experience and Apparitions*
6. God, if He exists, possess the attributes of omnipotence, omniscience, and omni-benevolence: *see Chapter 1*
7. God, if He exists, had a desire for divine incarnation: *see Chapter 8: The Resurrection of Jesus and Swinburne R. (2003[A]); Part 1, Ch. 2*

8. God, if He exists, had a desire for Jesus to be crucified and Resurrected: *see Chapter 8 and Swinburne R. (2003^A); Ch. 12*
9. God, if He exists, desires humans to be resurrected: *see Chapter 1: Omnibenevolence and Chapter 8: The Will of God*
10. There are naturalistic possibilities for human resurrection: *see Chapter 8: Naturalistic Possibilities*

These are the primary supportive auxiliary assumptions that are justified to some degree by reason and testimony, but there are also secondary ones embedded in the argument; for example, the free-will theodicy, cosmological assumptions concerning the universe. There also innocuous assumptions in background knowledge k as well, such as, the evidence of near-death experiences and apparitions are not deliberate hoaxes and a myriad of other assumptions.

Much of this independent support herein is not empirical evidence, but is based upon reasoned argument; although such assumptions may not be perfect auxiliaries, I believe they do, collectively, provide sufficient justification to claim that the likelihoods are well-defined.

It is mostly through the use of likelihood ratios that an assessment is made between the two positions; that is, $\Pr(e|h \,\&\, k$ (including all assumptions)) compared to $\Pr(e|\neg h \,\&\, k$ (including all assumptions)) where e is the evidence under consideration and h the relevant hypothesis/proposition.

It could be argued that the addition of multiple auxiliary assumptions increases the complexity of the likelihood and, through the principle of simplicity, lowers its probability. However, the use of likelihood ratios negates this danger as the complexity of each likelihood function is similar and it is the ratio value that works within Bayes' theorem; moreover, the principle of simplicity is questionable.[9]

The auxiliary assumptions do not appear as separate hypotheses within the relevant likelihood functions, but are assumed to reside within background knowledge k in those functions. They are embedded in the overall text, often with counter arguments and replies with the intention of the hypothetical members of the intersubjective group forming a view as to the degree of probabilistic force they have, combined with the main hypothesis, on the evidence. Thus, although the likelihood functions may be technically well-defined, having passed the test of genuine possibility, their justification will vary subjectively. For example, the auxiliary assumption within background knowledge k that God, if He exists, desired Jesus to be crucified and Resurrected will be, for some, questionable. As an assumption, it helps make the likelihood $\Pr(r|p \,\&\, R \,\&\, k)$ well-defined, where r is human resurrection, p the existence of God, R the Resurrection of Jesus, and k background knowledge in which, *inter alia*, this auxiliary assumption resides.[10] The degree to which that assumption is justified will undoubtedly vary as it did with the hypothetical intersubjective group; the reader will make their own subjective decision on this point.

NOTES

1. The thesis is a combination of Pierre Duhem's 1904/5 thesis and Willard Van Orman Quine's 1951 article "Two Dogmas of Empiricism."
2. Gillies D. (1993); pp. 112ff.
3. Sudduth M. (2016); p. 18.
4. Sober E. (2008); p. 142 for the use of auxiliary assumptions in likelihoods related to the intelligent design argument.
5. Sudduth M. (2016); Ch. 9.
6. Sturgeon S. (2000); Ch. 5.
7. Sudduth M. (2016); Ch. 9.
8. Sober E. (2003); p. 38.
9. Ch. 3; *Simplicity*.
10. Note that the auxiliary assumption in k, *God desires Jesus to be crucified and Resurrected* varies with $Pr(r|p \& R \& k)$ and $Pr(r|\neg p \& R \& k)$ as the latter function assumes God does not exist.

Appendix 2

Overdetermination of Evidence

There are additional evidential elements that other authors have addressed, such as mediumship, reincarnation, and poltergeists.[1] Although it is reasonable to consider the probabilistic relationship of these elements to the resurrection hypothesis individualistically through, for example, Bayesianism or likelihoodism, with a *cumulative* Bayesianism analysis there is a problem. With such a cumulative process, if one evidential element, resulting in a change in the degree of belief in a hypothesis, is followed by consideration of another evidential element that is roughly equivalent, the charge of double counting would apply. It is a case of overdetermination of evidence; that is, in terms of conditional probability, there is more evidence available than is necessary to confirm the hypothesis. Any one of the evidential elements, if true, is sufficient to confirm or disconfirm the relevant hypothesis to the same degree, but collectively not twice the value. For example, with NDEs, and mediumship, the experiencer frequently claims to be making personal contact with post-mortem persons. Indeed, the essential character of each of these processes is personal contact with the dead and it is that character that probabilistically confirms the resurrection hypothesis. Contingent characteristics, such as dormant brain activity in a NDE, or speaking in the voice of the deceased in the case of medium contact, is less relevant. Thus, testimonies of these experiences expressed propositionally, and taken as evidence of the afterlife, are not entirely distinct and to consider them in a cumulative Bayesian analysis would result in an exaggerated valuation.

The following example may clarify the problem of overdetermination of evidence in conditional probability:

> Let:
> h: Tom burgled the house.
> e_1: Dick saw Tom climbing through the house window.
> e_2: CCTV recorded Tom climbing through the house window.
> e_3: Harry photographed Tom climbing through the house window.
> BCON: the Bayesian conditionalization process.[2]
> k: background knowledge; left out for the purposes of clarity only.[3]

Through Bayes' theorem via likelihood ratios, the following cumulative progression emerges:[4]

1. $\Pr(e_1|h)/\Pr(e_1|\neg h) > 1$; $\Pr(h|e_1) > \Pr(h) = \text{BCON } \Pr(h)_{\text{NEW}}$
2. $\Pr(e_2|h)/\Pr(e_2|\neg h) > 1$; $\Pr(h_{\text{NEW}}|e_2) > \Pr(h)_{\text{NEW}} = \text{BCON } \Pr(h)_{\text{VERY NEW}}$
3. $\Pr(e_3|h)/\Pr(e_3|\neg h) > 1$; $\Pr(h_{\text{VERY NEW}}|e_3) > \Pr(\text{h})_{\text{VERY NEW}} = \text{BCON } \Pr(h)_{\text{VERY VERY NEW}}$

e_1, e_2, and e_3 each demonstrate that Tom climbed through the house window, but the truth of any one is sufficient to demonstrate this. Thus, a cumulative analysis, as above, would lead to over evaluating the probabilistic force of the evidence on the hypothesis.

The risk of over evaluating the probabilistic force of the evidence on the hypothesis in the above example can be demonstrated by consideration of background knowledge k, left out above for clarity only. Knowledge acquired from $\Pr(e_1|h \,\&\, k)$ concerning e_1 carries over to consideration of $\Pr(e_2|h \,\&\, k)$ as part of k, and expanding k in the function to demonstrate this, we have $\Pr(e_2|h \,\&\, e_1 \,\&\, k)$. The same applies to the negation of the hypothesis; that is, $\Pr(e_2|\neg h \,\&\, e_1 \,\&\, k)$ and given that e_1 is so similar to e_2, then $\Pr(e_2|h \,\&\, e_1 \,\&\, k) \approx \Pr(e_2|\neg h \,\&\, e_1 \,\&\, k)$ and therefore, $\Pr(e_2|h \,\&\, e_1 \,\&\, k)/\Pr(e_2|\neg h \,\&\, e_1 \,\&\, k) \approx 1$; this leads to little or no confirmation of h through Bayes' theorem using this likelihood ratio. Thus, 2 and 3 above are incorrect, but the potential for error for a cumulative Bayesian analysis remains.

Despite the problem of overdetermination of similar evidence, multiple witnesses plus CCTV recording can, *ceteris paribus*, probabilistically confirm h. Thus, the evidential proposition *CCTV recorded Tom climbing through the window and two witnesses testified to this* (e_4) is relevant evidence and captures the three evidential elements e_1, e_2, and e_3.[5] Thus, the following would then apply:

4. $\Pr(e_4|h)/\Pr(e_4|\neg h) > 1$; $\Pr(h|e_4) > \Pr(h) = \text{BCON } \Pr(h)_{\text{NEW}}$

In this case the value of $\Pr(h)$ has not been increased sequentially by considering the probabilistic force of h on e_1 then e_2 and then e_3. Moreover, the facts of e_1, e_2, and e_3, as multiple witness testimony and a CCTV recording, contribute to conditionalizing on e_4, resulting in BCON $\Pr(h)_{\mathrm{NEW}}$.

The probabilistic force of evidence in a justifiable cumulative case is when it approaches the hypothesis from different angles, such as stolen goods found at Tom's home, prior convictions for burglary and selling stolen goods as well as the testimonies and CCTV recording of Tom climbing through the window.

Similarly, NDEs and mediumship could be considered collectively in, for example, the proposition *there are testimonies of contact made with post-mortem persons by mediums and persons who claim to have had a NDE*. Evidence in support of the truth of this proposition, for the purposes of Bayesian conditionalization, could then be drawn from testimonial evidence of the two processes, providing a greater chance of conditionalization.

A parallel situation arises with testimony of apparitions and poltergeists in support of the resurrection hypothesis, where the essential character of the evidence is that of post-mortem persons contacting living persons.[6] To avoid exaggerated evaluation in a cumulative Bayesian analysis with each evidential element, a general evidential proposition could be considered, such as, *there are testimonies from living persons of post-mortem persons making contact with them in the form of apparitions and poltergeists*. Then, for purposes of Bayesian conditionalization, the Bayesian could look for evidence supporting the truth of the proposition via recorded data on apparition and poltergeist phenomena.

A similar situation arises with the Resurrection of Jesus and reincarnation, both being examples of human resurrection, albeit, in the former case, both human and divine. The same principle applies, namely, express both as a general evidential proposition, such as, *there is recorded testimony of the Resurrection and testimony of human reincarnation*. The evidence for both instances can then be employed for Bayesian conditionalization depending on the quality of the evidence.

The Bayesian may consider one evidential element to dominate an interrelated evidential set and, rather than conjoin them into one evidential proposition, select the dominant evidence for her analysis. Herein, I have limited myself to considering the Resurrection of Jesus, NDEs, and apparitions on the grounds of the quality and abundance of evidence. Moreover, each evidential element approaches the resurrection hypothesis from a different angle; namely, evidence of resurrected persons on Earth, evidence of living persons contacting post-mortem persons, and evidence of post-mortem persons contacting living persons.

NOTES

1. For example, Lester D. (2005); Braude S. E. (2003); Almeder R. (1992); Johnson R. C. (1989); Ducasse C. J. (1961).
2. Bayesian Conditionalization; Ch. 2.

3. Background Knowledge; Ch. 2.

4. Cumulative Bayesianism; Ch. 2.

5. It is important that the inclusive evidential proposition e is not worded in such a way as it becomes entailed by the hypothesis h, as $\Pr(e|h) = 1$; namely, maximum certainty.

6. Demonic interpretations of apparitions and poltergeists have little application to a human resurrection hypothesis.

Bibliography

Adams R. M. (1990)—*Middle Knowledge and the Problem of Evil*—in Adams M. M. & Adams R. M. (Eds) (1990); pp. 110–125.

Adams M. M. & Adams R. M. (Eds) (1990)—*The Problem of Evil*—Oxford: Oxford University Press.

Alexander E. (2012)—*Proof of Heaven: A Neurosurgeon's Journey into the Afterlife*—New York: Simon & Schuster.

Almeder R. (1992)—*Death and Personal Survival: The Evidence for Life After Death*—Maryland: Rowman & Littlefield.

Alston W. P. (1991)—*The Inductive Argument from Evil and the Human Cognitive Condition*—in Howard-Snyder D. (1996) (Ed.); pp. 97–125.

Alston W. P. (1993)—*Perceiving God: The Epistemology of Religious Experience*—Ithaca: Cornell University Press.

Ayer A. J. (1954)—*Freedom and Necessity* in Watson G. (1982) (Ed.); pp. 15–23.

Atemeyer R. and Hunsberger B. (1997)—*Amazing Conversions*—New York: Prometheus Books.

Aquinas T. [1274] (1991)—*Summa Theologiæ*—McDermott (Edt.); 1991; Texas: Christian Classics.

Augustine K. (2015)—*Near-Death Experiences are Hallucinations*—in Martin M. and Augustine K. (2015) (Eds); Ch. 25.

Baker L. R. (2000)—*Person and Bodies: A Constitution View*—Cambridge: Cambridge University Press.

Baker L. R. (2007)—*Persons and the Metaphysics of Resurrection* in Gasser G. (2010) (Ed.); pp. 161–176.

Barrow J. D. and F. J. Tipler (1996)—*The Anthropic Cosmological Principle*—Oxford: Oxford University Press [First published 1986].

Bartholomew D. J. (2001)—*Uncertain Belief—Is it Rational to be a Christian?*—Oxford: Clarendon Press [First published 1996].

Bell V. et al (2011)—*Correlates of perceptual distortions in clinical and non-clinical populations using the Cardiff Anomalous Perceptions Scale (CAPS): Associations with anxiety and depression and a re-validation using a representative population sample*—Psychiatry Research 189; pp. 451–457.

Birnbaum D. (1989)—*God and Evil: A Jewish Perspective*—New Jersey: Ktav Publishing House Inc.

Blackmore S. J. (1984)—*A Postal Survey of OBEs and Other Experiences*—in Journal of the Society for Psychical Research 52 (796); pp. 225–244.

Blumenfield D. (1995)—*Perfection and happiness in the best possible world*—in Jolley N. (1995); pp. 382–410.

Boethius A. M. S. $_{[C6th]}$ (1999)—*Consolation of Philosophy*—Fox S. (trans.); Ontario: In Parentheses Publications.

Borjigin J., Lee U.; Liu T, Pal D., Huff S., Klarr D, Sloboda J., Hernandez J., Wang M. M. Mashour G. A. (2013)—*Surge of neurophysiological coherence and connectivity in the dying brain*—in the Proceedings of the National Academy of Sciences of the United States of America 110; No. 35; pp. 14432–14437.

Bradley B., Feldman F. & Johansson J. (Eds.) (2012)—*The Oxford Handbook of Philosophy of Death*—Oxford: Oxford University Press.

Bradley R. D. (2015)—*Why Survival is Metaphysically Impossible*—in Martin M. and Augustine K. (2015) (Eds); Ch.11.

Braine D. (1998)—*Cosmological Arguments*—in Davies B. (1998); pp. 42–54.

Braude S. E. (2003)—*Immortal Remains: The Evidence for Life after Death*—Maryland: Rowman & Littlefield.

Burpo T. with Vincent L. (2010)—*Heaven is for Real*—Tennessee: W Publishing Group.

Carnap R. (1950)—*Logical Foundations of Probability*—Chicago: University of Chicago Press.

Caruso G. D. (2013) (Ed.)—*Exploring the Illusion of Free Will and Moral Responsibility*—Maryland: Lexington Books.

Cicero M. T. [45BC] (1896)—*De Natura Deorum*—(Trans. Brooks F.)—London: Methuen & Co.

Chisholm R. M (1964)—*Human Freedom and the Self*—in Watson G. (1982) (Ed.); pp. 24–35.

Chiu L. P. W. (1989)—*Differential Diagnosis and Management of Hallucinations*—Journal of the Hong Kong Medical Association 41; No. 3; pp. 292–297.

Clegg B. (2006)—*The God Effect: Quantum Entanglement, Science's Strangest Phenomenon*—New York: St Martin's Press.

Coleman D. (1988)—*Hume, Miracles and Lotteries*—Hume Studies XIV; No. 2 (Nov. 1988); pp. 328–346.

Colman S. and De Luccia F. (1980)—*Gravitational effects on and of vacuum decay*—Physical Review D. Vol. 21; Number 12; June 1980; pp. 3305–3315.

Corcoran K. (2001)—*Physical Persons and Postmortem Survival without Temporal Gaps*—in Corcoran K. (2001) (Ed.)—*Soul and Body Survival*—Cornell University Press; pp. 201–217.

Corcoran K. (2006)—*Rethinking Human Nature: A Christian Materialist Alternative to the Soul*—Grand Rapids, MI: Baker Academic.

Cortese F. (2013)—http://hplusmagazine.com/2013/06/17/clearing-up-misconceptions-about-mind-uploading/ (on 05–09–14 @ 13.00).

Cottingham J. (2015)—*How to Believe*—London: Bloomsbury.

Craig W. L. Smith Q. (1995)—*Theism, Atheism, and Big Bang Cosmology*—New York: Oxford University Press [First published 1993].

Dancy J. (2003)—in Singer P. (Ed.) (2003); pp. 411–420.

David F. N. (1962)—*Games, Gods and Gambling. The Origins and History and Probability and Statistical Ideas from the Earliest Times to the Newtonian Era*—London: Harper.

Davidson D. (1970)—*Mental Events*—in Rosenthal D. M. (1991); pp. 247–256.

Davis C. F. (1999)—*The Evidential Force of Religious Experience*—Oxford: Oxford University Press [First published 1989].

Davis N. (2003)—*Contemporary deontology*—in Singer P. (Ed.) (2003); pp. 205–218.

Davies B. (1998) (Ed.)—*Philosophy of Religion: a guide to the subject*—Washington: Georgetown University Press.

Davies M. (1995)—*The Philosophy of Mind*—in Grayling A. C. (1998); pp. 250–335.

Dawid P. and Gillies D. (1989)—*A Bayesian Analysis of Hume's Argument Concerning Miracles*—The Philosophical Quarterly 39; No 154; Jan. 1989.

Dennett D. C. (1993)—*Consciousness Explained*—London: Penguin Books [First published 1991].

Descartes R [1637] (2008)—*Discourse on Method*—Kennington R. (Trans.); 2008; Massachusetts: Focus Publishing/R Pullins Company.

Dilley F. B. (1977)—*Fool-proof proofs of God?*—in International Journal for Philosophy of Religion Vol. 8; No. 1, (1977); pp. 18–35.

Drange T. M. (2015)—*Conceptual Problems Confronting a Totally Disembodied Afterlife*—in Martin M. and Augustine K. (2015) (Eds); Ch. 12.

Draper P. (1996)—*Pain and Pleasure: An Evidential Problem for Theists*—in Howard-Snyder D. (1996) (Ed.); pp. 12–29 [First published 1989].

Ducasse C. J. (1961)—*The Belief in a Life After Death*—Illinois: Charles C. Thomas Publisher.

Durkheim E. (1961)—*The Elementary Forms of the Religious Life*—New York: Collier Books; Trans. Swain J. [First published 1915].

Earman J. (1996)—*Bayes or Bust*—Massachusetts: MIT [First published 1992].

Evans G. (2000) (Ed. McDowell J.)—*The Varieties of Reference*—Oxford: Clarendon Press [First published 1982].

Evans H. (2002)—*Seeing Ghosts: Experiences of the Paranormal*—John Murray Publishers Limited.

Everitt N. (2004)—*The Non-existence of God*—London: Routledge.

Fischer J. M. and Mitchell-Yellin B. (2016)—*Near-Death Experiences: Understanding Visions of the Afterlife*—New York: Oxford University Press.

Flew A. (1950)—*Theology and Falsification*—in 'University' (an undergraduate journal at University College of North Staffordshire) reprinted in MacIntyre A. and Flew A. (Eds.) (1955)—*New Essays in Philosophical Theology*; pp. 96–99; London: SCM Press.

Flew A. (1968)—*God and Philosophy*—London: Hutchinson & Co Limited [First published 1966].

Foster J. (1996)—*The Immaterial Self*—Oxford: Routledge [First published 1991].

Frankfurt H. G. (1971)—*Freedom of the Will and the Concept of a Person* in Watson G. (1982).

Freud S. (1927)—*The Future of an Illusion*—translated and reprinted in Strachley J. (Ed.) (2001); pp. 1–56.

Gale R. M. (1996)—*Some Difficulties in Theistic Treatments of Evil*—in Howard-Snyder D. (1996) (Ed.); pp. 206–218.

Gale R. M. and Pruss A. R. (1999)—*A New Cosmological Argument*—in Gale R. M. and Pruss A. R. (2003); pp. 365–380.

Gale R. M. and Pruss A. R. (2003)—*The Existence of God*—The International Research Library of Philosophy (Book 28); Burlington: Ashgate Publishing Ltd.

Gasser G. (2010) (Ed.)—Personal Identity and Resurrection: How Do We Survive Our Death?—UK: Ashgate Publishing Limited.

Gellman J. I. (1997)—*Experience of God and the Rationality of Theistic Belief*—Ithaca: Cornell University Press.

Gerard J. Hughes SJ (1998)—*Omniscience*—in Davies B. (1998) (Ed.).

Gillies D. (1991)—*A Bayesian Proof of a Humean Principle*—British Journal of the Philosophy of Science 42; pp. 255–256.

Gillies (1991[A])—*Intersubjective Probability and Confirmation Theory*—British Journal of the Philosophy of Science 42; pp. 513–533.

Gillies D. (1993)—*Philosophy of Science in the Twentieth Century: four central themes*—Oxford: Blackwell Publishers.

Gillies D. (2003)—*Philosophical Theories of Probability*—London: Routledge [First published 2000].

Glymour C. (1980)—*Theory and Evidence*—Princeton: Princeton University Press.

Grayling A. C. (1998)—*Philosophy 1*—New York: Oxford University Press.

Greene B. (2011)—*The Hidden Reality*—London: Allen Lane.

Greenfield S. A. (2000)—*The Private Life of the Brain*—London; Allen Lane: The Penguin Press.

Griffith-Dickson G. (2000)—*Human and Divine: an introduction to the philosophy of religious experience*—London: Gerald Duckworth & Co Ltd.

Griffith-Dickson G. (2005)—*The Philosophy of Religion*—London: SCM Press.

Guthrie S. (1993)—*Faces in the Clouds*—Oxford: Oxford University Press.

Habermas G. R. and Licona M. R. (2004)—*The Case for the Resurrection of Jesus*—MI: Kregel Publications.

Hacking I. (2001)—*An Introduction to Probability and Inductive Logic*—Cambridge: Cambridge University Press.

Halpern P. (2012)—*Edge of the Universe: A Voyage to the Cosmic Horizon and Beyond*—New Jersey: John Wiley & Sons Inc.

Haraldson E., Godmundsdottir A., Ragnarsson A., Loftsson J., Jonsson S. (1977)—*National Survey of Psychical Experiences and Attitudes Towards the Paranormal in Iceland*—in Research in Parapsychology; pp. 182–186.

Hardy A. (1979)—*The Spiritual Nature of Man*—Oxford: Clarendon Press.

Hasker W. (1992)—*The Necessity of Gratuitous Evil*—in Faith and Philosophy 9; No1 (Jan'92); pp. 23–44.

Hauskeller M. (2016)—*Death and Meaning*—to be published in a collection on transhumanism and the philosophy of immortality—Came D.; Burwood S.; Ornella A. (Eds.)—Oxford: Oxford University Press. Also: https://www.academia.edu/12561637/Death_and_Meaning (18/12/2015).

Hay D. and Morisy A. (1978)—*Reports of Esctatic, Paranormal, or Religious Experience in Great Britain and the United States: A Comparison of Trends*—in Journal for the Scientific Study of Religion 17; No. 3 (Sep., 1978); pp. 255–268.

Hay D. (1979)—*Religious Experience Amongst a Group of Post-Graduate Students: A Qualitative Study*—in Journal for the Scientific Study of Religion 18; No. 2 (Jun., 1979); pp. 164–182.

Hay D. (1982)—*Exploring inner space: Is God still possible in the twentieth century*—UK: Penguin Books.

Hawking S. & Mlodinow L. (2010)—*The Grand Design*—London: Bantam Press.

Hick J. (1964) (Ed.)—*The Existence of God*—New York: Macmillan.

Hick J. (1967)—*Faith and Knowledge*—London: Macmillan and Company Limited [First published 1957].

Hick J. (1974)—*Evil and the God of Love*—Fontana Library [First published 1966].

Hick J. (1976)—*Death and Eternal Life*—UK: William Collins Sons & Co Ltd Hick J. (1981)—*Soul-Making Theodicy*—in Rowe W. L. (2001)(Ed.); pp. 265–281.

Hobbes T, (1662)—*An Answer to Bishop Bramhall's Book, called "The Catching of the Leviathan"* in The English Works of Thomas Hobbs of Malmesbury—London: John Bohn (1840).

Hoffman J. & Rosenkrantz G. S. (2002)—*The Divine Attributes*—Oxford: Blackwell Publishers.

Howard-Snyder D. (1996) (Ed.)—*The Evidential Argument from Evil*—Indiana: Indiana University Press.

Howard-Snyder D., Bergmann M. & Rowe W. L. (2001)—*An Exchange on the Problem of Evil*—in Rowe W. L. (2001)(Ed.); pp. 124–158.

Howson C. (2000)—*Hume's Problem: Induction and the Justification of Belief*—Oxford: Clarendon Press.

Howson C. (2011)—*Objecting to God*—Cambridge: Cambridge University Press.

Howson C. & Urbach P. (1991)—*Scientific Reasoning: The Bayesian Approach*—Illinois: Open Court Publishing Company [First published 1989].

Hume D. [1748] (1993); Steinberg E. (Ed.)—*An Enquiry Concerning Human Understanding*—Indianapolis: Hackett Publishing Company.

Hume D. [1757] (2010)—*The Natural History of Religion*—Kila: Kessinger Publishing.

Hume D. [1779] (1998); Popkin R. H. (Ed.)—*Dialogues Concerning Natural Religion*—Indianapolis: Hackett Publishing Company.

Herculano-Houzel S. (2009)—*The Human Brain in Numbers: A Linearly Scaled-up Primate Brain*—Frontiers Human Neuroscience 3: 31.

Inwagen P. van (1975)—*The Incompatibility of Free Will and Determinism*—in Watson G. (1982); pp. 46–58.

Inwagen P. van (1978)—*The Possibility of Resurrection*—International Journal for the Philosophy of Religion; 9; pp. 114–121.

Inwagen P. van and Zimmerman D. (2007) (Eds.)—*Persons: Human and Divine*—Oxford: Oxford University Press.

Ipsos MORI (1998)—*Paramormal Survey*—https://www.ipsos-mori.com/research publications/researcharchive/2082/Paranormal-Survey.aspx (December 2016).

Ipsos MORI (2007)—*Survey on Beliefs*—https://www.ipsos-mori.com/research publications/researcharchive/246/Survey-on-Beliefs.aspx (December 2016).

Ipsos Social Research Institute (2011)—*Supreme Being(s), the Afterlife and Evolution* -http://www.ipsos-na.com/news-polls/pressrelease.aspx?id=5217 (December 2016).

James W. (1985)—*The Varieties of Religious Experience*—Marty M. E. (Ed.); New York: Penguin Books [First published 1902].

Jantzen G. M. (1984)—*God's World God's Body*—London: Darton, Longman and Todd.

Johnson R. C. (1967)—*Nurslings of Immortality*—London: Hodder and Stroughton [First published 1957].

Johnson R. C. (1989)—*The Imprisoned Splendour*—London: Pelegrin Trust [First published 1953].

Kant I. 1781 (1929)—*Critique of Pure Reason*—Smith N. K. (Trans.)—London: McMillan Press Ltd.

Keynes J. M. (1996)—*Treatise on Probability*—London: Macmillan Press Ltd [First published 1921].

Kim J. (1984)—*Epiphenomenal and Supervenient Causation*—in Rosenthal D. M. (Ed.) (1991); pp. 257–265.

Kim J. (2007)—*Causation and Mental Causation*—in McLaughlin B. P. and Cohen J. (2007) (Eds.); pp. 228–242.

Kneale W. and Kneale M. (1988)—*The Development of Logic*—Oxford: Clarendon Press [First published 1962].

Koch C. (2012)—*Consciousness: confessions of a romantic reductionist*—Massachusetts: MIT.

Koene R. A. (2012)—*Experimental Research in Whole Brain Emulation: The Need for Innovative In-Vivo Measurement Techniques*—International Journal of Machine Consciousness Vol. 4, No. 1 (2012).

Koons R. C. (1997)—*A New Look at the Cosmological Argument*—in American Philosophical Quarterly; 34(2).

Kuhn T. S. (2012)—*The Structure of Scientific Revolutions*—Chicago: University of Chicago Press [First published 1962].

Kyburg H. E. Smokler & H. E. (eds.) (1980)—*Studies in Subjective Probability*—New York: Robert E. Krieger Publishing Company.

Lambert Y. (2004)—*A Turning Point in Religious Evolution*—Journal of Contemporary Religion 19; 1; pp. 29–45.

Laplace (Maquis de) P. S. (1998)—*A Philosophical Essay on Probabilities*—New York: Sringer; translated from the 5th French Edition of 1825 by A. I. Dale [First published as Essai philosophique sur les probabilités (1814)—Paris: Gauthier-Villars].

Leibniz G. W. (1710) (1951)—*Theodicy*—London: Routledge & Kagan Paul Ltd; Huggard E. M. (Trans.).

Leibniz G. W. [1696] (1973)—*On the Principle of Indiscernibles*—in Leibniz Philosophical Writings; Parkinson G. H. R. (Ed.;Trans.) and Morris M. (Trans.)—London: J. M. Dent & Sons; pp. 133–135.

Lennox J. C. (2009)—*God's Undertaker: Has Science Buried God?*—Oxford: Lion Hudson.

Lester D. (2005)—*Is There Life After Death: An Examination of the Empirical Evidence*—North Carolina: McFarland & Company.

Lewis D. (1986)—*Counterfactuals*—Oxford: Basil Blackwell [First published 1973].

Libet B., Gleason C. A., E. W. Wright and D. K. Pearl (1983)—*Time of conscious intention to act in relation to onset of cerebral activity (readiness potential). The unconscious initiation of a freely voluntary act*—in Brain; 106; pp. 623–642.

Livio M. (2009)—*Is God a Mathematician?*—New York: Simon & Schuster.

Llinás R. and Churchland P. S. (1996) (Eds.)—*The Mind-Brain Continuum*—Massachusetts: MIT.

Llinás R. and Paré D. (1996)—*The Brain as a Closed System Modulated by the Senses*—in Llinás R. and Churchland P. S. (1996); Cht. 1.

Loads A. and Rue L. D. (1993)—*Contemporary Classics in Philosophy of Religion*—Illinois: Open Court Publishing Company [First published 1991].

Lock J. [1632–1704] (1974)—*An Essay Concerning Human Understanding*—London: J M Dent & Sons Ltd.

Lommel P. van P.; Wees R. van; Meyers V.; Elfferich I. (2001)—*Near-death experience in survivors of cardiac arrest: a prospective study in the Netherlands*—in The Lancet 358; Dec. 15 (2001); pp. 2039–2045.

Lommel van P. (2010)—*Consciousness Beyond Life: The Science of the Near-Death Experience*—New York: Harper Collins Publishers.

Long J. with Perry P. (2010)—*Evidence of the Afterlife: The Science of Near-Death Experiences*—New York: Harper Collins.

Loose J. (2013)—*The Metaphysics of Constitution and Accounts of the Resurrection*—Philosophy Compass 8/9; (2013); pp. 857–865.

Mackie D. (1999)—*Personal Identity and Dead People*—Philosophical Studies; 95; pp. 219–242.

Mackie J. L. (1955)—*Evil and Omnipotence*—in Mind 64; pp. 200–2012; reprinted in Adams M. M & Adams R. M. (Eds.) (1990); pp. 25–37.

Mackie J. L. (1982)—*The Miracle of Theism*—Oxford: Clarendon Press.

Maimonides M. [1190] (1956)—*The guide for the perplexed*—Friedländer M. (trans.); New York: Dover Publications Inc.

Manson N. A. (2003) (Ed.)—*God and Design: The teleological argument and modern science*—New York: Routledge.

Marsh M. N. (2010)—*Out-of-Body and Near-Death Experiences: Brain-State Phenomena or Glimpses of Immortality*—Oxford: Oxford University Press.

Martin M. and Augustine K. (2015) (Eds);—*The Myth of an Afterlife: The Case Against Life After Death*—Maryland: Rowman & Littlefield.

Maslin K. T. (2007)—*An Introduction to the Philosophy of Mind*—Cambridge: Polity Press.

McLaughlin B. P. and Cohen J. (2007) (Eds.)—*Contemporary Debates in Philosophy of Mind*—Oxford: Blackwell Publishing Ltd.

McGrew, Timothy, "Miracles," *The Stanford Encyclopedia of Philosophy* (Winter 2015 Edition), Edward N. Zalta (ed.), URL = <http://plato.stanford.edu/archives/win2015/entries/miracles/>.

Meierding L. (1998)—*The Consensus gentium Argument*—Faith and Philosophy15; 3; July, 1998.

Mesle C. R. (1991)—*John Hick's Theodicy: A Process Humanist Critique*—London: Macmillan Press Limited.

Miller B. (1996)—*A Most Unlikely God*—University of Notre Dame Press.

Moody R. A. (2001)—*Life After Life*—London: Rider [First published 1975].

Morse M. and Perry P. (1990)—*Closer to the Light: Leaning from Children's Near-Death Experiences*—New York: Random House.

Mosteller F. & Youtz C. (1990)—*Quantifying Probabilistic Expressions*—in Statistical Science 5; pp. 2–34.

NatCen (2008)—*Belief in Religious Miracles*—British Religion in Numbers; June 2008; http://www.brin.ac.uk/figures/#ChangingBelief 7. 39.

Noonan H. W. (2003)—*Personal Identity*—Oxford: Routledge [First published 1989].

Nozick R. (1984)—*Philosophical Explanations*—Oxford: Oxford University Press [First published 1981].

O'Connor D. (1998)—*God and Inscrutable Evil: in defense of theism and atheism*—Maryland: Rowman & Littlefield.

O'Connor T. and Jacobs J. D. (2010)—*Emergent Individuals and the Resurrection*—European Journal for the Philosophy of Religion; 2; pp. 69–88.

Olson E. (1999)—*The Human Animal: Personal Identity without Psychology*—Oxford: Oxfornd University Press [First published 1997].

Olson E. (2007)—*What Are We? A Study in Personal Ontology*—New York: Oxford University Press.

Olson E. (2012)—*The Person and the Corpse*—in Bradley B., Feldman F. & Johansson J. (Eds.) (2012); pp. 80–96.

Olson E. (2010)—*Immanent Causation and Life After Death*—in Gasser G. (2010) (Ed.); pp. 51–66.

Otte R. (2002)—*Rowe's Probabilistic Argument from Evil*—Faith and Philosophy 19; 2 (April 2002).

Owens J., Cook E. W. and Stevenson I. (1990)—*Features of "near-death experience" in relation to whether or not patients were near death*—in Lancet 336; No 8724; pp. 1175–1177.

Paley W. [1794] (1871)—*A View of the Evidences of Christianity*—London: Christian Evidence Committee of the Society for Promoting Christian Knowledge.

Paley W. [1802] (2008)—*Natural Theology—Eddy M. D. & Knight D. (Eds.); Oxford: Oxford University Press.*

Palmer J. and Dennis M. (1974)—*Spontaneous Case Trends: A Community Mail Survey of Psychic Experiences*—in Research in Parapsychology; pp. 130–133.

Parfit D. (1987)—*Reasons and Persons*—Oxford: Clarendon Press [First published 1984].

Paterson R. W. K. (1995)—*Philosophy and the Belief in a Life After Death*—London: MacMillan Press Limited.

Pew Survey (2010)—*Religion Among the Millennials*—http://www.pewforum.org/2010/02/17/religion-among-the-millennials/.

Philipse H. (2014)—*God in the Age of Science*—Oxford: Oxford University Press.

Plantinga A. (1981)—*Is Belief in God Properly Basic?*—in Loads A. and Rue L. D. (1993); pp. 93–105.

Plantinga A. (1982)—*The Nature of Necessity*—Oxford: Clarendon Press [First published 1974].

Plantinga A. (1988)—*Epistemic Probability and Evil*—in Archivio di filosofia 56; reprinted in Howard-Snyder D. (1996); pp. 69–96.

Plantinga A. (1990)—*God and other minds: a study of the rational justification of belief in God*—Ithaca, NY: Cornell University Press [First published 1967].

Plantinga A. (1993)—*Warrant and Proper Function*—Oxford: Oxford University Press.

Plantinga A. (1996)—*God, Freedom and Evil*—Michigan: William B. Eerdmans Publishing Company [First published 1974].

Plantinga A. (1998)—*Degenerative Evidence and Rowe's New Evidential Argument from* Evil—in NOÛS 32:4; pp. 531–544.

Plantinga A. (2007)—*Materialism and Christian Belief*—in Inwagen P. van and Zimmerman D. (2007) (Eds.); pp. 99–141.

Portugal A. C. (2003)—*Probability Theories and the Justification of Theism—in* Pre-Proceedings of the 26th International Wittgenstein Symposium. Austrian Ludwig Wittgenstein Society, Kirchberg am Wechsel, pp. 283–285.

Puhle A. and Parker A. (2017)—*An Exploratory Study of Lucid Dreams Concerning Deceased Persons*—Journal of the Society for Psychical Research; Vol. 81; No. 3; (July 2017).

Ramsey F. P. (1926)—*Truth and Probability*—The Foundations of Mathematics and other Logical Essays; R. B. Braithwaite (ed); the Humanities Press—New York (1950); reprinted in H. E. Kyburg & H. E. Smokler (eds.); (1980); pp. 23–52.

Rees M. (2003)—*Other Universes: A Scientific Perspective*—in Manson N. A. (2003).

Rosenthal D. M. (1991) (Ed.)—*The Nature of Mind*—New York: Oxford University Press.

Rovelli C. (2016)—*Reality Is Not What It Seems: The Journey to Quantum Gravity*—UK: Allen Lane.

Russell B. (1956)—*Logic and Knowledge*—London: Unwin Hyman Ltd [*On Denoting* First published 1905].

Russell B. and Copleston F. (1964)—*Debate on the Existence of God*—in Hick J. (ed.) (1964); pp. 167–191.

Romer C. J. (2011)—*Why Everything We Think We Know About Ghosts Is (Probably!) Wrong*—Anomaly 45; (Nov. 2011); pp. 93–121.

Rowe W. L. (1991)—*Ruminations about Evil*—in *Philosophical Perspectives* 5; pp. 69–88.

Rowe W. L. (1996)[A]—*The Problem of Evil and Some Varieties of Atheism*—in Howard-Snyder D. (1996) (Ed.); pp. 1–11 [First published in American Philosophical Quarterly 16; (1979)].

Rowe W. L. (1996)[B]—*The Evidential Argument from Evil: A Second Look*—in Howard-Snyder D. (1996) (Ed.); pp. 262–285.

Rowe W. L. (2001)(Ed.)—*God and the Problem of Evil*—Massachusetts: Blackwell Publishers Inc.

Sacks O. (2012)—*Hallucinations*—London: Picador.

Sainsbury M. (2001)—*Logical Forms: An Introduction to Philosophical Logic*—Oxford: Blackwell Publishers Limited.

Sandberg A. (2013)—*Feasibility of Whole Brain Emulation*—in Muller V. C. (Ed.) *Theory and Philosophy or Artificial Intelligence*—Berlin: Springer; pp. 251–264.

Sandberg A. and Bostrom N. (2008)—*Whole Brain Emulation: A Roadmap, Technical Report #2008–3*—Future of Humanity Institute, Oxford University.

Satori P. (2014)—*The Wisdom of Near-Death Experiences*—London: Watkins Publishing.

Schiffer S. (2003)—*The Things We Mean*—Oxford: Oxford University Press.

Shoemaker S. (1970)—*Persons and their pasts*—in Shoemaker S. (1984); Ch. 2.

Shoemaker S. (1984)—*Identity cause and mind*—Cambridge: Cambridge University Press.

Sidgwick H., Johnson A., Myers F. W. H., Podmore F., Sidgwick E. M. (1894)—*Report on the Census of Hallucinations*—in Proceedings of the Society for Psychical Research 10; pp. 25–422.

Singer P. (Ed.) (2003)—*A Companion to Ethics*—Oxford: Blackwell Publishing [First published 1991].

Slors M. (2004)—*The Closest Continuer View Revisited*—Croatian Journal of Philosophy; 12.

Smith R. (2013)—*A Century of Apparitions: Revisiting the Census of Hallucinations in the 21st century*—PhD thesis; Coventry University UK.

Sobel J. H. (2004)—*Logic and Theism: Arguments for and Against Belief in God*—Cambridge: Cambridge University Press.

Sober E. (2002)—*Bayesianism—its scope and limits*—in Swinburne R. (2002) (Ed.); p. 21.

Sober E. (2003)—*The Design Argument*—in Manson N. (2003); pp. 27–54.

Sober E. (2008)—*Evidence and Evolution: the logic behind the science*—Cambridge: Cambridge University Press.

Sober E. (2009)—*Absence of evidence and evidence of absence: evidential transitivity in connection with fossils, fishing, fine-tuning, and firing squads* in Philosophical Studies 143:63–90.

Soon C. S. et al (2008)—*Unconscious determinants of free decisions in the human brain*—Nature Neuroscience; Vol. 11; No. 5; May 2008.

Strachley J. (Ed.) (2001)—*The Standard Edition of the Complete Works of Sigmund Freud*—Vol: XXI in collaboration with Freud A. and assisted by Strachley A. and Tyson A.—London: Vintage.

Starbuck E. D. (1897)—*A study of Conversion*—in The American Journal of Psychology 8; No. 2 (Jan., 1897), pp. 268–308.

Strawson G. (2008)—*Real Materialism: and other Essays*—Oxford: Clarendon Press.

Strawson P. F. (1966)—*Self, Mind and Body* in Rosenthal D. M. (1991); pp. 58–62.

Steinhart E. C. (2014)—*Your Digital Afterlife: Computational Theories of Life after Death*—UK: Palgrave Macmillan.

Sturgeon S. (2000)—*Matters of Mind: Consciousness, reason and nature*—London: Routledge.

Sudduth M. (2016)—*A Philosophical Critique of Empirical Arguments for Postmortem Survival*—New York: Palgrave Macmillan.

Sullivan A., Voas D. and Brown M. (2012)—*The art of asking questions about religion*—Centre for Longitudinal Studies; Institute of Education; University of London (Nov. 2012).

Swinburne R. (1970)—*The Concept of Miracle*—London: MacMillan and Company.

Swinburne R. (1973)—*An Introduction to Confirmation Theory*—London: Methuen.

Swinburne R. (1977)—*The Coherence of Theism*—Oxford: Clarendon Press.

Swinburne R. (1997)—*Simplicity as Evidence of Truth*—Milwaukee: Marquette University Press.

Swinburne R. (1998)—*Providence and the Problem of Evil*—Oxford: Oxford University Press.

Swinburne R. (2001)—*Epistemic Justification*—Oxford: Clarendon Press.

Swinburne R. (2002) (Ed.)—*Bayes's Theorem*—London: for The British Academy by Oxford University Press.

Swinburne R (2003)—*The Argument to God from fine-tuning*—in Manson N. A. (Ed); pp. 105–123.

Swinburne R (2003[A])—*The Resurrection of God Incarnate*—Oxford: Clarendon Press.

Swinburne R. (2004)—*The Existence of God*—Oxford: Clarendon Press [First publish 1979: Clarendon].

Tien A. Y. (1991)—*Distributions of Hallucinations in the Population*—in Social Psychiatry and Psychiatric Epidemology 26; pp. 287–292.

Tilly T. W. (2000)—*The Evils of Theodicy*—Oregon: Wipf and Stock Publishers.

Tippler F. J. (1994)—*The Physics of Immortality*—New York: Doubleday.

Tooley M. (2015)—*The Problem of Evil—The Stanford Encyclopedia of Philosophy* (Fall 2015 Edition), Edward N. Zalta (ed.), URL = <http://plato.stanford.edu/archives/fall2015/entries/evil/>.

Tyrrell G. N. M. (1973)—*Apparitions*—London: The Society for Psychical Research [First published 1943].

Unwin S. D. (2003)—*The Probability of God*—New York: Crown Forum.

Vargus M. (2013)—*If Free Will Doesn't Exist, Neither Does Water*—in Caruso G. D. (2013) (Ed.); pp. 177–202.

Vardy P. (1995)—*The Puzzle of God*—London: Fount [First published 1990].

Wainwright W. J. (1999)—*Philosophy of Religion*—California: Wadsworth Publishing.

Watson G. (1982) (Ed.)—*Free Will*—Oxford: Oxford Readings in Philosophy.

West D. J. (1948)—*Mass-Observation Questionnaire on Hallucinations*—in Journal of the Society for Psychical Research 34; pp. 184–196.

West D. J. (1990)—*A Pilot Census of Hallucinations*—in Proceedings of the Society for Psychical Research 57 (215); pp. 163–207.

West D. J. (1995)—*Notes on a Recent Psychic Survey*—in Journal of the Society for Psychical Research 60 (838); pp. 168–171.

Wiggins D. (1980)—*Sameness and Substance*—Oxford: Blackwell.

Williams B. (1973)—The Makropulos case: reflections on the tedium of immortality in *Problems of the Self*—Cambridge: Cambridge University Press; pp. 82–100.

Wiseman R. (2015)—*Paranormality*—London: Pan Books [First published 2011].

Wykstra S. J. (1990)—*The Humean Obstacle to Evidential Arguments from Suffering: On Avoiding the Evils of Appearance*—in Adams M. M. & Adams R. M. (Eds) (1990); pp. 138–160 [First published 1984].

Wykstra S. J. (1996)—*Rowe's Noseeum Arguments from Evil*—in Howard-Snyder D. (1996) (Ed.); pp. 126–150.

Yandell K. (1994)—*The Epistemology of Religious Experience*—New York: Cambridge University Press [First published 1993].

YouGov (2011)—*University of Hertfordshire Ghosts*—http://cdn.yougov.com/today_uk_import/Results_for_University_of_Hertfordshire_Ghosts_OM_2011_025_03_02_11.pdf (December 2016).

Zimmerman D. (1997)—*Immanent Causation*—Philosophical Perspectives 11; pp. 433–471.

Zimmerman D. (2010)—*Bodily Resurrection: The Falling Elevator Model Revisited*—in Gasser G. (2010) (Ed.); pp. 33–50.

Index

About the Author

Dr William Hunt has been an educationist for most of his life, first as a school teacher at William Ellis School in North London and then as Principal of GSM London, a college of higher education in South East London offering bachelors, masters, and doctorate degree program in the domain of management, law, and computer science. He taught Statistics for the Department of Extra Mural Studies of the University of London and completed his PhD in the philosophy of religion at Heythrop College, University of London in 2007 and returned in 2015 as a visiting Research Fellow. He is a member of the British Society for the Philosophy of Religion and the Alister Hardy Trust. His other interests include membership of the Sherlock Holmes Society of London, he is a keen skier and enjoys rugby football, alas only as a spectator these days. He lives in Blackheath, South East London, and is married with five children.

CPSIA information can be obtained
at www.ICGtesting.com
Printed in the USA
BVOW03*1743131117
499399BV00007B/12/P

9 781498 526067